AF540817

UNDERSTANDING ETHOLOGY

UNDERSTANDING ETHOLOGY

By

Ashok Kumar

Dept. of Zoology
Bundelkhand University
Campus Department
Jhansi

DISCOVERY PUBLISHING HOUSE PVT. LTD.
NEW DELHI-110 002

First Published-2010

ISBN 978-81-8356-506-6

Published by:

DISCOVERY PUBLISHING HOUSE PVT. LTD.

4831/24, Ansari Road, Prahlad Street,
Darya Ganj, New Delhi-110002 (India)
Phone: 23279245 • Fax: 91-11-23253475
E-mail: parul.wasan@gmail.com
info@discoverypublishinggroup.com
Website: www.discoverypublishinggroup.com

Printed at:

Sachin Printers
Delhi

Preface

The present title "Understanding Ethology" has been written for those students interested in careers in diverse fields of biological sciences. It provides a structured approach to learning by covering all the important topics in a uniform, systematic format. The book has been comprehensively designed incorporating recent advances in this fast moving field. It also provides accessible information on ethology in compact form for undergraduate students in biology and related life sciences. It is intelligible to the educated layman, though it deals with some complex ideas. It is an adequate text for all the requirements of students in this area. In addition, busy lecturers who require a quick reference compendium will find it useful, particularly for tutional planning. Simple, yet hopefully clear figures and tables are provided throughout the book.

The over-riding goal of this book, and indeed of the whole *Understanding series*, is to present the essential information concering ethology in a compact, readily accessible form which leads itself to student learning and revision. The convergence of various approaches has generated a rich panorama of detail, the significance of which we are still attempting to unraval. The present text has been written as an introduction to this rapidly growing field.

To make the work more comprehensive and informative, the author has consulted many authoritative books, research journals, abstracts, monographs etc., so there can be no claim to originality except in the manner of treatment.

The author expresses his thanks to his friends and colleagues whose continue inspirations have initiated him to bring out this book.

The author expresses his gratitude to Mr. Wasan and staff of M/s Discovery Publishing House Pvt. Ltd. for their whole hearted co-operation in the publication of this book.

In the mean time, the author will remain sincerely responsible for any shortcomings of the book and be grateful to the readers for their suggestions and constructive criticism for the continuous betterment of the book. He takes this opportunity to appeal to the readers to send their suggestions straightaway to his Publisher.

Author

CONTENTS

1

INTRODUCTION

The behaviour of animals is flexible and diverse. During a typical day an individual animal may perform, in a particular order, dozens of different behaviour patterns, or units. A male bird in the spring, for example, may sing at dawn, and then feed (using many different foraging techniques), hide from predators, defend its territory, and court a female, during one day. Ethologists have not progressed far in describing the mechanisms that control the complete behavioural repertory of an animal through its life—that is, the way it decides what behaviour patterns to perform in a given set of circumstances, how to trade off its motives to perform one activity rather than another, and by what physiological mechanisms these decisions are made. If ethologists have not progressed far in solving the total problem, they have made many important and curious discoveries about the mechanisms controlling some individual behaviour patterns. This is particularly true of animal senses. The general, elemental mechanisms controlling behaviour are known, even if how they are combined to produce the full behavioural output of an animal is not. Muscular contractions are an important mechanical source of many behaviour patterns. Animals control their muscles by their nervous systems.

The nervous system also carries information from the sense organs and, by mechanisms of excitation and inhibition among its different parts, integrates the animals's senses with its internal tendencies and preferences. But the study of animal nervous systems has not in all cases turned out to be the easiest route to understanding how they control their behaviour, and we shall consider some examples of how animal senses have been successfully studied at a higher level than that of sensory neurons, and their behavioural choices at a higher level than the influence of neurons on one another.

The Spider's Web

Most spiders are solitary, with one individual living an each web a wondering as solitary hunters. Among spiders that build sheet like webs–the kinds so after seen on lawns and in hedges– Here are species in which the spider lings remain as the web and are fed by their another. In some species the spider lings build their own small webs next to their mothers. The final stage in the evolution of sociality is communal construction of the web. Several spiders may run out to subdue a large prey. The nature of these social systems strongly suggests that they evolved via the familial route. Among orb-weaving spiders, however, soci lity appears to have volved via the parasocial route. Individual build their own webs in most species.

The start of sociality was probably clumping of webs in unusually favourable sites. This led in a few cases to the joining of webs and construction of communal retreat. Each spider builds and occupies her own web and there is no communal subduing of prey. There is no care of the young, overlaping of generations, or division of labor. In the social orb weavers there is no evidence that the females are closely related to one another. Perhaps the araneid spiders, such as the common garden spider *Araneus diadematus*, which do build orb webs, have a concept of the web; we cannot tell. However, observations of spiders in action suggest that they follow a series of rules, which in themselves would be sufficient to lead the spider to build a web even if it had no idea of what the end point should be.

Not all the flexible details of the spider's building program have been worked out, but the main rules were discovered simply by watching them build. The observations are easy to make. Orb webs are spun by common spiders all over the world; and the whole operation takes less than half an hour. The end product, the orb web, is a regular structure made up of frame, radial spokes, and catching spiral. The stages of its construction were first formalized by *Hans Peters* in 1939, and they are as follows. The spider starts with the frame and spokes. To begin with, she spins a thread with one unattached end and allows it to be blown by the breeze. The other end is attached to the spider, by the spinnerets of the abdomen, which are the source of the thread. The loose end will soon become entangled in some object, such as a nearby twig. The spider then bites through her end of the thread and, having attached a new thread at her point of departure, walks of down

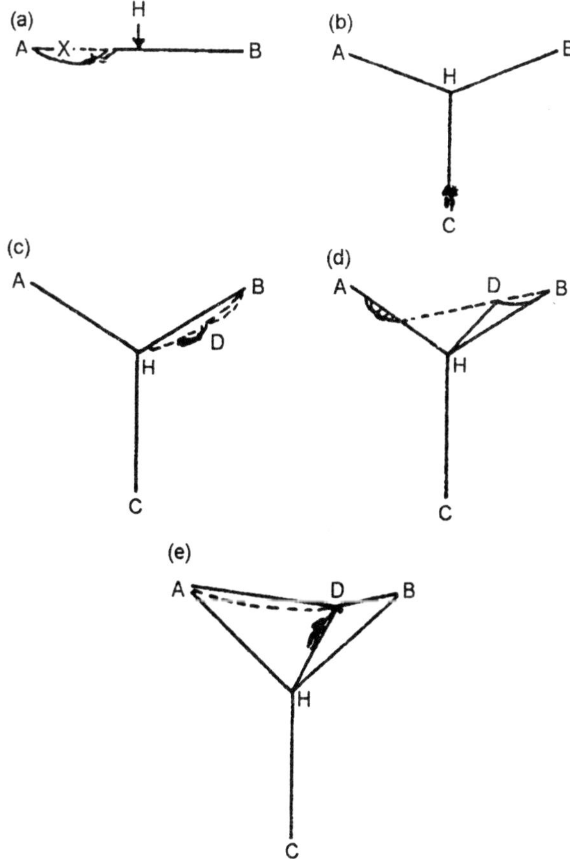

Fig. 1.1. The stage in the construction of an orb web.

the wind-cast thread, spinning another thread as she goes. When she arrives at the other end, she attaches the new thread. She then turns, walks some distance up the thread and attaches a new thread there. She now allows herself to fall, spinning a thread behind her, to a third attachment point (C). The Y-shaped structure provides the scaffolding for the web. She next builds by turns the radial spokes and frame. She returns to the centre H (which stands for hubs, as the centre of the completed web is called), attaches a thread, walks to the outside (B) spinning a thread as she goes, and attaches the other end of if at B; she then walks back down the thread to a, point D where she attaches a new thread. Now, after she has walked on to A, she has constructed a frame from B to A and a radial spoke from D to H. The same sequence of movements final web the angles of the radii at the hub are rather constant, at about 15.'

The radii are built in a regular order, the next spoke always being added in the largest unfilled sector of the web. In other words, the spider measures the angles between the existing spokes at the hub, and chooses to spin the next spoke between the two spokes subtending the largest angle. One she has returned to the hub after building a spoke she pulls each spoke with her front legs, which may be how she measures the angles, perhaps by estimating the tension in each spoke. At all events, sight is unnecessary for correct building because spiders can build normal webs in complete darkness, or when sightless. Following the rule for where to spin the next spoke results in spokes being laid down in the kind of order illustrated in Figure, with each new spoke in a different sector from the previous one. Once the frame and radii are complete (The formation of frame and radii takes about five minutes), the spider starts on the spiral. She builds it in two stages.

Starting from the hub, she first spirals outwards, to lay down an '*auxiliary spiral.*' She does this by walking outwards, is an arithmetic

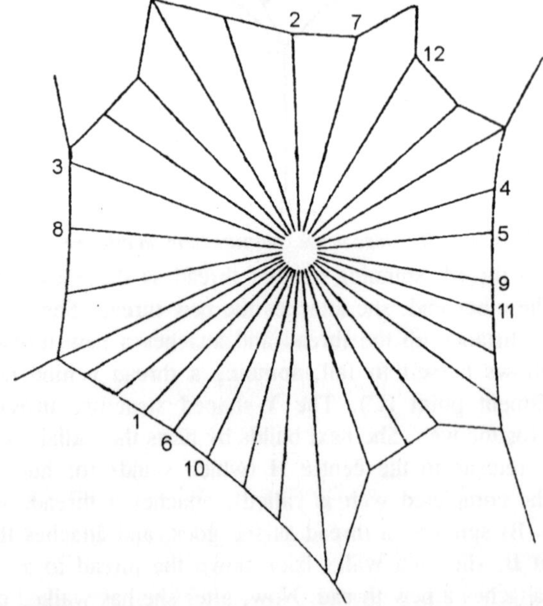

Fig. 1.2. An orb web, with most of its radial spokes, but no catching spiral. The numbers indicate the order in which the radii are laid down.

spiral, spinning and attaching a thread behind her, until she reaches the edge. The auxiliary spiral is a temporary structure. For at the outside she turns round and, working inwards now, spins the sticky she cuts the auxiliary spiral, the function of which was to act as a guideline for laying down the catching spiral. The whole web is finished once the sticky catching thread has spiralled into the hub. Actually, the spiral is not a perfect spiral; it is asymmetrical. Orb webs are not typically circular. They are elongated in the bottom half, with the hub above the centre. The bottom is lengthened by means of a ladder of sticky switch-backs in the bottom sector of the web, which can be seen in Figure. Only after a few turns of the ladder does the spider begin the true spiral. The main rules of the orb web spider are first to build a Y-shaped scaffold, then, in a set order, the frame and radial spokes; and finally the auxiliary and sticky spirals.

Each main rule contains a set of subrules for measuring angles and walking set distances up certain threads. By following the program of elementary rules the spider can-build a complex structure without having a plan of it in her head. This is not to say that the spider lacks such a plan. Spiders surely are more complex machines than this account has admitted. There is probably more to the control of web building than the series of rules considered here, But this fact does not principle which the spider's web illustrates: that apparently complex structures can be build up from a fairly simple set of rules. The principles is of general importance. It is often possible to break down the apparent complexity of animal behaviour by analysing it into a series of 'rules of thumb' that the animal itself could, in some form, be following. Searching for the rules is a powerful method of studying the control of behaviour, because it is easier to pursue research on simple rules than one complex ones.

It is easier to follow up the nervous control of a leg movement than of conceptual thought. Of course, complex systems can be studied too; but because it is more difficult it simply takes longer. More sophisticated rules are most easily studies if they are added piecemeal on to a concrete foundation of knowledge. We therefore, advance fastest in our understanding of behaviour if we search for the simplest hypothetical mechanisms. Ethologists do not invoke less simple mechanisms—of conceptual thought or conscious calculation, for example—if they are not needed to explain the behaviour that is being observed. So far we have thought of an animal as executing a series of rules. But this execution must in its turn by controlled by some mechanisms. The controlling mechanism ultimately is the nervous system; and to see how that works we must change our study animal, from the araneid spider, to squids and crickets.

The Nervous System

It is the integrated system and is composed of a network of nerve fibres. A nerve fibre is are bundle of cells called *neurons*. A muscle contracts when the neuron attached to it, called a motor neuron, becomes electrically active. Neurophysiologists first worked out how neurons work in one particular kind of neuron which is found in the squid. When a squid senses that an enemy is nearby, it suddenly contrasts its body, forcing water out of the area called its *mantle cavity*, and jets away. The squid may also discharge a screen of ink to cover its escape.

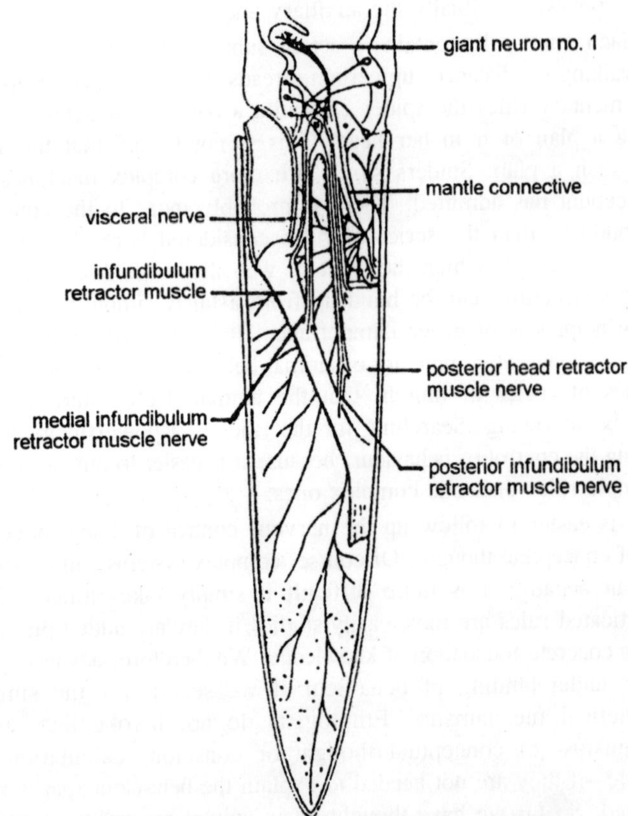

Fig. 1.3. The course of the giant neurons of the squid.

The 'escape reaction', as it is called, is controlled by a large (and therefore easily studied) neuron, which extends in stages from just outside the squid's brain at one end of its body, to the muscles, the contraction of which at the other end of the body effects the escape reaction. The impulse travel in the nerve fibre in the electrochemical means. The neurosome is polarized at rest and has positively not as well as negatively charged ions on different sides of the membrane. When the neuron is activated, positively charged sodium ions rush into the neuron. The inside of the neuron in the activated area becomes electrically positive relative to the outside. This pulse of depolarization, as it is called, travels rapidly along the neuron until it reaches the muscle. The electric current then stimulates the muscle to contract. But what stimulated the motor neuron to begin with? We must turn to the other end of the neuron, away from the muscle, where the motor neuron is joined to many other neurons at a junction called a *synapse*. Here is the origin of motor neuron activity.

Chemical neurotransmitters are released by the neuron at the synapse, and if sufficient is released, the motor neuron will fire into action. The motor neuron is thus controlled by the activity of the many neurons preceding it in the nervous system, whose activity is in turn controlled by the central nervous system in the brain. Sensory input from the external environment is also carried to the brain encoded in electrical depolarizations of neurons, called *sensory neurons*, which pass from sensory organ to the brain. Speaking very generally, the behaviour of an animal is mainly directed by its brain, which integrates the actions of thousands of neurons, but by processes that are too little understood to be worth dwelling upon. The neuronal control of some simple behaviour patterns is, however, quite well understood. The singing of crickets is an example, and it will serve to illustrate how the nervous system produces behaviour. Singing in crickets is not a response to an external stimulus. It is influenced by external factors such as temperature; but is not a direct response to them. It is caused by the endogenous action of the nervous system. Male crickets sing, by scraping their wings together, to attract females.

The movement of wings is controlled by a set of muscles in the thorax (the thorax is the middle section of the insect, between the head and the abdomen). Motor neurons run from the thoracic ganglion (a ganglion is a mini-brain, an aggregation of many nerves) to the singing muscles. It takes many muscle fibres to move a wing, and each muscle fibre has its own motor neuron. All the muscle fibres must contract at the same time, and the co-ordination is produced by a 'command interneuron', onto which all the motor neurons join. Soon after the command interneuron fires, all the motor neurons fire in unison.

The co-ordinatory function of the command interneuron is situated fires, all the motor neurons fire in unison. The co-ordinatory function of the command neuron has been proved by experiment. The command interneuron fires, all the motor neurons fire in unison. The co-ordinatory function of the command neuron has been proved by experiment. The command interneuron is situated in the cricket's neck. A neurophysiologist can locate it and join it to an electricity supply. If the electricity is switched on, all the motor neurons fire together, and the cricket produces a perfect song. It will produce its song even if it has had its head cut off. The command neuron, motor neurons, muscles, and wings form a complete singing machine.

Sense Organs or Receptors

All animals live a fluctuating environment and they need to be able to find their way around their environment, find food, recognize the species and sex of other individuals, even (in some cases) whether another individual is a member of the same, or a different, group, and detect the signals sent to them. For all these functions they rely on their sense organs. Different kinds of animals have different sets of sense organs. The set of sense organs possessed by each kind of animal is appropriate to the environment in which it lives. For example, species of fish and shrimp which live in dark, underground caves do not have eyes, or have eyes so reduced that they no longer work; there is no advantage in having light-sensitive organs where there is no light. The human set of eyes, ears, touch, and relatively poor taste and smell is just one particular, not very common, set of sense organs; most species of animals live in a world of senses very different from our. The various receptors of animals can be divided into three groups: *exteroceptors*, *enteroceptors* and *proprioceptors*. This division was suggested by the neurophysiologist *Charles Sherrington* at the beginning of the century. Exteroceptors sense the state of the environment outside the animal; enteroceptors the state of the animal's body; proprioceptors the animal's movement by sensing the position of its muscles.

In mammals, enteroceptors include organs that sense the body temperature and chemoceptors that sense the concentrations of chemicals such as hormones and carbon dioxide. The classic five human senses—sight, hearing, touch, taste, smell—are effected by exteroceptors. In fact there is no perfect classification of these external senses. Taste and smell are both chemical senses (as are many of the enteroceptors). Hearing and touch are both mechanical senses. Other species possess other senses which do not easily fit into the five-way division. But, bearing in mind that the classification is crude, we can consider quickly some examples of each kind.

Galvano Receptors

The electrical sense is lacking in humans but studied in fishes where it is greatly developed. Their electrical sense is not the same as the sensitivity to pain by which we become aware of an electric shock, but another sense, comparable to our sense of hearing or smell. Some kinds of fish, such as dogfish (which is a member of the elasmobranch group that includes sharks and rays), use their electric sense to find food buried in the bottom sand by the disturbance to the electric field that the buried living matter causes. The electric sense organs of elasmobranch fish are called the *ampullae of Lorenzini* and are a set of jelly-filled tubes beneath the skin. Other animals, such as a honey-bees, and bacteria, can sense magnetism. Experiments we shall meet later suggest that at least some kinds of birds can also sense the magnetic field.

Chemoreceptors

Chemical sense organs are found in many place in different kinds of animals. The sea hare, which is a favourite animal for work on the nervous system, can smell seaweed, on which it lives. By recording the activity of the nervous system at different parts of the animal while it is smelling seaweed, it has been found that the chemical sense organs are in its tentacles. Houseflies have chemical receptors in their feet, which enables them to detect food (such as sugary water, in an experiment) by walking into it. Most insects have chemical receptors in their antennae.

Tangoreceptors

The antennae also contain mechanical sense organs; but an insect's whole surface has mechanical sense organs on it. Mechanical sense organs all work by means of tiny hairs. When the hair is bent a sensory neuron attached to it fires into action. Hearing works as a mechanical sense, and is also effected by small movement-sensitive hairs connected in some way (depending on the species) to a membrane that is set in oscillation by sound. Fish and some amphibians possess special organs for detecting water pressure called the *lateral line organ*.

The lateral line is a channel under the skin of each side of the animal, with little holes leading to the outside. The flow of water into the lateral line allows the fish to measure to movement of water with respect to itself.

Photoreceptors

The final class of sense organs is the light receptors. Eyes are found is more or less complex form in many kinds of animals. But eyes are not the only light-sensitive organ known in nature. Insects, for instance, have three light sensitive ocelli on the top of their heads, behind their compound eyes. The functions of the ocelli are uncertain.

Echolocation in Bats

Bats are the aerial mammals showing volant adaptations. There are over 900 species of bats. The group has perhaps proliferated because its echolocatory sense allows it to live in an environment containing almost no competitors. They fly by night, feeding on night-flying insects, particularly moths. Bats can fly with ease in complete darkness; they do not collide with obstructions, and they catch their prey on the wing, They are capable of flying around a laboratory room which is criss-crossed with a network of wires of a diameter of as little as 3 hundredths of an inch, and catching flying moths from a range of 8 feet. How do bats achieve this? The greatest experimental biologist of the eighteenth century, the Italian *Lazzaro Spallanzani,* could not solve the problem. He did find that if he stuffed up the bats ears their ability to avoid obstructions declined. But he did not know how to explain this result.

Bats remained a puzzle until after advanced equipment of recording and producing sound has been developed along with radar in the Second World War. The equipment was applied to the bat puzzle by *Donald Griffin* in the 1950s. He proved that the little brown bat finds its way around by listening to the echos of high-pitched sounds that it makes itself. Dogs can hear sounds of the higher frequency of about 40,000 Herfa. Humans cannot hear sounds much outside the frequency range 2000-20,000 Hertz (Hz). (The higher the frequency the higher the pitch). Bats make sounds mainly of 20,000 Hz and higher—some emit sounds of up to 100,000 Hz—and most bat sounds are therefore inaudible to humans. One advantage to the bat of using such high-pitched sounds is that it is undisturbed by background noise (most noises in nature have a frequency lower than 20,000 Hz). By concentrating on high-frequency sounds, bats live in a world silent except for their own noises.

It is essential to the bat that there should be no interfering background noise. In an experiment, *Griffin* blasted noises of frequencies of more than 20,000 Hz into a room with obstructions. Flying bats now bumped into the obstacles and fell into the floor. As well as showing the importance of a silent background, this experiment also gives part of the evidence that bats use high frequency sounds to find their way around. Echolocation works by measuring the time interval between releasing a short pulse of sound and hearing its echo. The longer it takes the echo to come back, the further away the object must be. The sound pulses have to be very short, to prevent overlap between the emitted sound and its echo: that bat can hear itself as well as the echo. The pulses of sound are indeed very short, the little brown bat releases four or five distinct pulses every second during ordinary flight. It increases the rate of pulses when it detects on object, and as it flies closer to the object its pulse rate rises further. It does so because, as it approaches the object, the delay until the echo comes back becomes shorter and shorter. It therefore, makes it pulses shorter, again to prevent overlap between pulse and echo. Echolocation becomes ineffective beyond distances of about 30 40 metres, because sound is rapidly absorbed in air. But even for these short distances the bat must listen for a very faint echo of its much louder original pulse; the sound pulse emitted by the bat is 2000 times louder than the echo. The bat therefore has the problem that it has to make a loud noise and then hear a soft noise immediately afterwards. When an ear has been blasted with a loud sound it becomes less sensitive to recover after listening to a very loud noise. Bats have a number of methods of solving this problem. One is to make the ear less sensitive when emitting the sound.

A nerve going to the muscle of the ear is automatically activated whenever the bat releases its sound pulse. This nerve causes the bat's ear to relax about 5 thousands this of the a second before the pulse is emitted, and to recover about 10 thousandths of a second later. The ear will then be a peak sensitivity for picking up the echo. Bats are not the only kind of animal to use echolocation. They are however, by far the most extensively studied. Other animals which echolocate are the dolphin and other 'toothed' whales, small mammals called shrews, and a bird which lives in dark caves called the Malayan cave swiflet.

Use of Sensory Informations

Animal behaviour is not only a set of simple responses to various environment situations; but we can make some points about how an animal uses its sensory information without implying that it is all there is to behaviour. All sense organs send their knowledge in coded form to the central nervous system by means of sensory neuron. Different sensory neurons are stimulated by different properties of the environment. Light sensitive neurons, for instance, contain a light sensitive pigment, which, when illuminated, changes its chemical form and causes the neuron to burst into action. A special neuron is required to sense light: if you shone light on any other kind of neuron it would have no effect on it; only a neuron containing a light sensitive pigment is set in action by light. Most behaviour patterns are controlled in the central nervous system which integrates the information from the sense with other neuron systems, to control the animal's behavioural output.

Some behaviour patterns are not centrally controlled. They are called peripheral reflexes (of which the human knee jerk is an example), and are controlled by a simple system of one sensory neuron and one motor neuron. The sensory neuron connects directly on the motor neuron. The sensory neuron connects directly on to the motor neuron, which in turn controls the muscles that effect the behaviour pattern. When the sensory neuron fires (after mechanical stimulation in the case of knee jerk reflex), it stimulates the motor neuron, which causes certain muscles to contrast.

We shall meet some other examples of reflexes; but it most behaviour patterns the sensory input and behavioural output of the animal are less directly connected, and the senses exert their influence on behaviour through the central nervous system. There is potentially an enormous amount of information about the environment, far more than an animal could make use of in deciding on a course of action. The environment, for instance, is brim-full of electromagnetic rays: it has a constantly changing pattern of light and dark, different colours, and a continual hum of x-rays and radio waves, together with high and low frequency radiation of which we are not normally aware.

Most of the information is, for any given behavioural decision of the animal, completely irrelevant. To avoid walking into a tree you only need to know where its edges are; the fine detail of colour patterns on the trunk and leaves do not matter. To behave appropriately, therefore, an animal has to select its information, and that is exactly what it does. Some of the selection is done in the sense organ itself, which will only sense certain patterns; and the rest of the selection is carried out as the information is centrally integrated into the nervous system. Given that animals are selecting from the available information in their environment, how can we find out what they are actually responding to? One method is neurophysiology. We can record the activity of sensory neurons when an animal is successively presented with a variety of objects.

We shall meet one example of this kind of study in the prey-catching behaviour of the toad. The same kind of question can also be tackled by a grosser method. We can ignore the physiological intermediary details, and find out what kind of environmental stimuli the animal responds to at the behavioural level. The simplest kind of sensory information is used in the responses called '*kinesis*' and '*taxis*'. In a kinetic response, the animal alters its rate of movement, in a random direction, according to the intensity of the stimulus. When the stimulus, which might be light, or moisture, is of the right intensity it slows down and thus spends more of its time under those conditions. Woodlice show a kinetic response to moisture. They move faster where it is drier, and therefore spend more time where it is moist.

The unicellular organism called *Paramecium* shows a kinetic response with respect to the local concentration of carbon dioxide. A taxic response, however, is directional. Negative phototaxis, for example, means that the animal moves away from light, as in fact does the maggot of the bluebottle fly. The taxic response is made by sensing the direction of the light, which is achieved by different techniques in different species. It requires only the most elementary kind of sensory information. The animal does not need to known anything about the light source, only that it is light and where it is coming from. Paramecia normally swim about engulfing food particles at random. They feed by using the cilia around their mouth to sweep bacteria and other food particles into their food vacuoles. If a group of paramecia are placed in a drop of water, they will soon gather around and feed upon any bacterial colonies present in the drop. This action implies that their feeding is not at random. For some reason they remain quite stationary when they come upon a colony of bacteria. Other experiments have shown that they will also collect around a variety of materials–cloth, wool and cotton. Their behaviour, at least in this case, appears to be in response to a certain class of objects and is not directed towards a specific goal. We shall see further examples of this kind of behaviour in multicellular organisms.

Paramecia sometimes group together even if there are no clumps of material. How would you account for this action? Experiments have shown that they will congregate near areas that are slightly acid.

Some of these experiments are illustrated. Apparently, carbon dioxide released during their own respiration upon them, survived, and continued to multiply. Those paramecia which did not possess this behavioural characteristic would not be so apt to survive and leave descendants. The type of behaviour we have just described can usually be explained in simple chemical or physical terms. It is called stereotyped behaviour a stereotype being anything that is fixed, unvarying, and showing little or no individuality. Such protist behaviour is not only stereotyped; it is also innate, or inherited. The paramecium did not learn that an acid solution means food. It positive response to an acid condition was innate. Few behaviour patterns, except perhaps chemical responses to food (or pheromones, are guided simply by the intensity of one stimulus.

More of the behaviour patterns that we can see animals performing are controlled also by the pattern of the stimulus in the environment. For example, in the case of light, the exact distribution of light and exact distribution of light and shade, which defines the shape of the image, would be important, rather than only the presence or absence of light. Animals must be able to recognize patterns in the environment, gulls do likewise. But, what exactly is it that stimulates the behaviour? How does a gull recognize an egg? Gerard Baerends and his colleagues made model eggs, which varied in their size, stippling, shape and colour. They presented herring gulls, on their nests, with choices of different model eggs, and recorded which ones were preferentially retrieved.

The gulls preferentially retrieve larger models, even if their size exceeds the natural size of an egg, and the sizes of an egg can therefore be used as a scale of comparison for the other variables, Baerends tested each model egg by giving gulls a choice between the model egg (which might be varied in its colour, shape or stippling) and a range of size of control models (of normal shape, colour and stippling). The gulls take more notice of some variables than of others, They take little notice of shape: an oblong model is as likely to be retrieved as an egg-shaped model of the same size (compare a with R). Nor do the fulls prefer eggs of natural colouration: they preferentially retrieve a green eggs of natural colouration: they preferential retrieve a green egg rather than a naturally coloured egg of the same size (compare c with R); but they are sensitive to colour, as they prefer green to brown eggs (compare b with c). The other variable, besides size, which strongly influences the Gulls' preference is stippling; they prefer stippled to unstippled eggs (compare d with c). It is as if natural

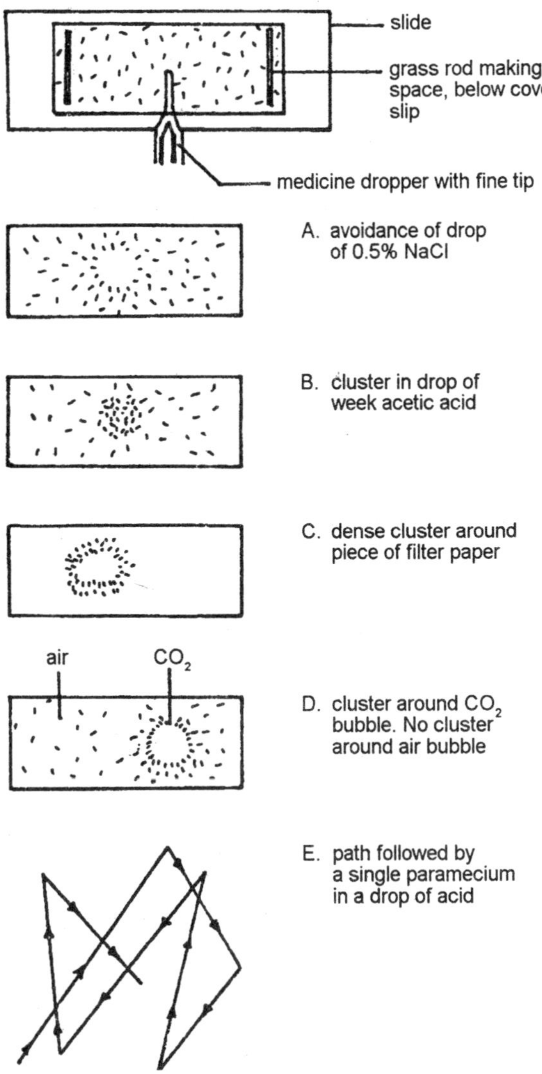

Fig. 1.4. Behavioural experiments with the paramecium.

colour and shape are not part of a gull's idea of an egg, but size and stippling are. They do not simply prefer to retrieve eggs of natural appearance, and indeed prefer to retrieve eggs of natural appearance, and indeed prefer larger than normal eggs, in which case the models take on the condition of a supernormal stimulus'; the model is a more attractive stimulus as the stimulatory property is exaggerated.

Fig. 1.5. A black-headed gull with a choice of an artificially large egg (of natural colour and stippling) and a natural egg. The gull is retrieving the large egg.

The supernormal stimulus in a sense deceives the egg recognition mechanism of the gull. In nature, the mechanism work to distinguish eggs from other objects; but it can-be tricked by experiment. However, than main conclusion from Baerends' experiment is that herring gulls recognize eggs mainly by the criteria of size and stippling. '*Behavioural assays*', such as the egg retrieval response of birds, not only reveal what stimulus patterns is recognized by the animal; they are also a revelatory method of studying the sensory powers of animals. If an animal can be shown, by appropriately controlled experiments, to behave in response to some property of the environment, it must be able to sense it. Karl von Frisch applied the method to demonstrate the hearing ability and colour sensitivity of fish. In that case, the physiologist von Hess had asserted that fish are colour blind and deaf; von Frisch doubted the assertion, and he successfully trained minnows to distinguish colours by rewarding them with food, and catfish to come out of a tube when he below a whistle. In both experiments he used a behavioural response to discover a sensory ability.

Behaviour Patterns

The behavioural output of an animal emerges as a sequence of many different behaviour patterns, and each change in the sequence can be through of as a behavioural 'choice' made by the animal. The question is how animals make those choices. Some will be responses to changed environmental stimuli. New sensory information is one factor causing an animal to choose one behaviour pattern rather than another, but it is not the only one. The same animal may not respond

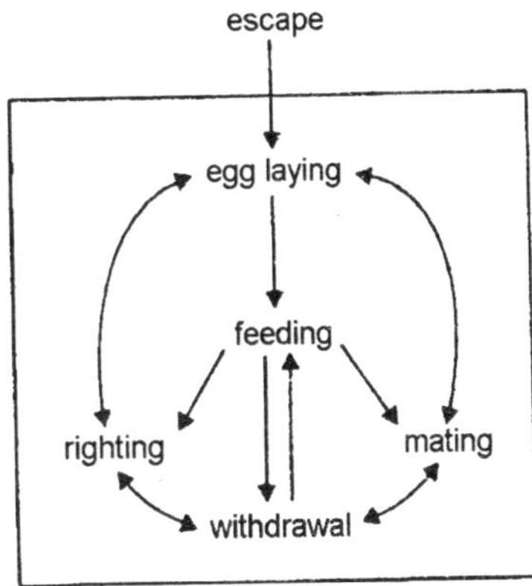

Fig. 1.6. The behavioural priorities of the intertidal snail pleurobranchia.

to the same stimulus in the same way on different occasions, and it may change what it is doing even while the environment appears to be constant. There are two, related reasons why an animal may behave differently when under similar environmental conditions. One is that its internal tendency to behave in a certain way may change. For example, when food is presented, it will become less likely to feed as it grown less hungry.

The other reason is the interaction of behavioural preferences; an animal may stop feeding in order to avoid a predator. The two reasons are related because the internal tendency of an animal to behave in a certain way is presumably determined by a balancing act among the consequences of all its possible responses to a given environment. The balancing act, and the resultant behavioural preferences or tendencies, are called the motivation of the animal. Let us consider how the motivation of an animal influences its behavioural choices. The behaviour in the freshwater fish called guppies (*Poecilia reticulata*), provides a clear example of the interaction of motivation and external stimulus.

The behaviour in question is the courtship of females by males. *Baerends* recognized three different behaviour patterns that a male may perform when courting a female: 'posturing' in front of a female, limited sigmoid movement, and a full sigmoid display. How does a male decide which to perform? The answer seems to depend on the size of the female and the male's own motivation to court, which can be independently measured by his colouration. Figure depicts the combinations of these two factors necessary for a male to court a female in each of the three ways.

The particular shape of the graphs is not important here; they are only to illustrate a general point, which is that, for any behaviour pattern in any species there will be some such graph of motivational tendency and external stimulus which describes the conditions under which it is performed. The guppy illustrates choice among different behaviour patterns of one class, courtship. What of interactions among different kinds of behavioural goal? Here we need a new example, which (unlike the courtship of guppies) is understood neurophysiologically. Actually, little progress has been made in the neurophysiological study of behavioural choices. Nervous analysis is difficult enough for single behaviour units, let alone interactions among many activities.

The study of the gastropod *Pleurobranchia* by *J. W. Davis* and his colleagues has, however, partly uncovered the neurophysiological control of six behaviour patterns. The six are feeding, egg laying, escape, withdrawal of the oral veil, righting, and mating. Take first the interaction of feeding and egg laying *Pleurobranchia* is a carnivorous snail which includes eggs in its diet. When a *Pleurobranchia* lays its own eggs, it switches off its feeding habit. Another behaviour pattern, escaping from predators, is performed in preference to all other activities. A fourth behaviour patter is to withdraw its oral veil on being touched. A snail with its oral veil withdrawn cannot feed, and its tendency to withdraw its veil interacts with its tendency to feed. If food is abundant, or the snail is not hungry, withdrawal has priority over feeding, and *vice versa* when food is scarce and the snail is hungry.

The other two activities studies by *Davis* are *righting* (turning the right way up) and mating. Having established the behavioural priorities by observation, he proceeded to their neurophysiology. The system has not been completely elucidated, but it is now known for instance that two neurons are responsible for inhabiting the 'withdrawal' response when a *Pleurobranchia* is feeding, and that the inhibition of feeding during egg laying is effected hormonally. The priorities of *Pleurobranchia*, by the way, do make sense, for without them it would eat its own eggs after laying them; and if escaping from danger did not have absolute priority, it would not survive to exercise its other behavioural preferences. Hormonal influences on behaviour can be divided roughly in two categories: educational and organizational effects.

In *educational effects* hormones act as triggering influences on the expression an informance of behavioural patterns. The *organizational* effects of hormones are manifested during an organism's development. The following classification will give a clear understanding of different types of effects (in behavioural aspects).

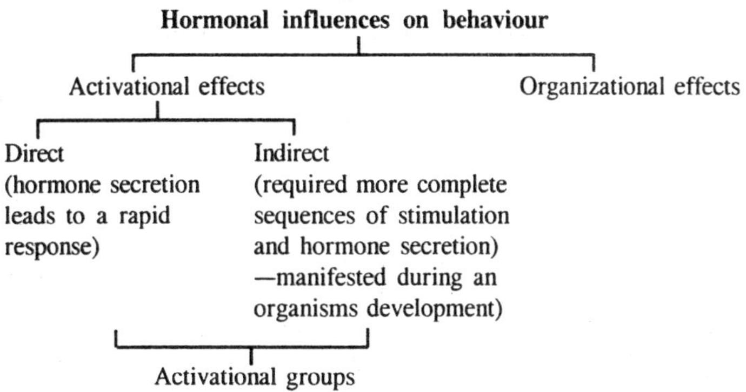

1. Sexual attraction
2. Eclosion
3. Development phrases
4. Moulting
5. Colour Change
6. Aggression and sexual behaviour
7. Secondary sexual characteristics

Interrelationship between Hormones and Behaviour

Investigations have used several technique to explore the between hormones and behaviour. These include:

1. *Radioimmunoassay* to directly measure circulating levels of a hormone though the use of immunological methods.
2. *Blood transfusion* to transfer the "hormonal state" of one animal at another in order to observe the behavioural effect
3. *Exterpation or removal, of a particular endocrine gland* to asses the absence of a specific hormone on behaviour.
4. *Bioassays* to indirectly assess circulating hormones levels by measuring a secondary characteristics such as skin gland, that is dependant on a particular hormone.
5. *Autoradiography* to localize the sites at which hormone uptake occurs.

6. *Hormone replacement therapy* injection of specific hormone or transplantation of a gland.

ACTIVATIONAL EFFECTS

Sexual Attraction

In certain species of cockroaches and moths the female release *pheromones* which act as sex attractants for male.

Example. If in a female cockroach *corpus allatum* is surgically removed after moulting the female will not produce pheromone after it becomes sexually mature. But if corpus allatum of another adult female is transplanted to it then it again starts producing pheromones.

Now a question arises that a why has this system of *sex attraction evolved*? There are several answers:

1. Some pheromones may serve to synchronize the reproductive activities of the two sexes of a given species.
2. Attractant using that a male will find a female and mate with her.
3. Series as a sexual excitatory function by bringing with members logically.
4. Emission of a species sex-attractant may be species isolating mechanism. It avoids gamete wastage.

Eclosion

The process whereby the adult form of an insect emerges from the pupa after metamorphosis is called *eclosion* and is another activational effect controlled hormonally. Many moth species enclose at species *specific line* of adry. The eclosion hormone which is produced by neurosecretory cells in the brain plays a critical role in this process.

Example. If the eclosion hormone is cirficted into pupa that are near the end of the metamorphosis eclosion behaviour, such as abdomen movements and being spreading after emergence can be activated at any time of the deny. Moths that have their brains removed usually emerge; therefore, the presence of eclosion hormone is not an absolute requirement for eclosion take place. The process, however, is not a as coordinated in brainless subjects and some activities (e.g. wing spreading)are usually absent. Thus although the hormone may not be necessary for eclosion, it does appear to be necessary for proper coordination of the sequence.

Development Phases

In adult male desert locusts sexual behaviour is exhibited when corpora allata are removed and when corpora allata is transplanted from other locust is restores its sexual behaviour, however, similar experiments have revealed that corpora allata are not needed for sexual behaviour in certain grasshoppers.

Moulting

In some types of norms and molluscs and uncrustaceous in investors have concentrated the presence of one or more neurosecretory or endocrine glands whose secretions effect several differentiation and malivation of gaindes and ethaulati reproduction. Many crustacean moult periodically as they grind. Removal of both eyestalk in these animals shortens the interval between, moults. If these are given the extracts of particular clearly, the gland produces a moult-inhibiting factor.

Colour Change

In short tailed weasels which undergoes seasonal changes in pelage, orcont during spring and fall melts. In spring *metamorphose stimulating hormone* (MSH) secretion increases and new brown hairs replace white coat color during the full months. MSH secretion is inhibited by the action of another hormone *melatonin* secretion by the sexual gland, the hair then is not argumented and returns to white.

Aggression and Sexual Behaviour

Example. When male ring doves are *castrated* they show decreased levels of aggressions courtship and copulation behaviour when treated with crystalline *testosterone* the normal levels of behaviours restored (*Banfield* 1971) *Mutchusin* (1969, 1971, 1978). Testosterone effects both on sexual and aggressive behaviour.

Secondary Sexual Characteristics

Example (i) The characteristics cock's becomes greatly decreased in castrated roosters. (ii) Male cats, which spray urine probably as a marking behaviour, often cease to spray after their testes are removed.

ORGANIZATIONAL EFFECTS

Example. Neonatal male rats pups were injected with estrogen. Histological examinations of the rats revealed some degeneration of the seminiferous tubles, where sperms are produced. The investigations mated that although these males showed mounting behaviour was irregular- the mounts were often incorrectly oriented and no ejaculation occurred. The *organizational effects* of hormones on female behaviour have been conducted on guinea pigs and Rhesus monkey. Female progency of females that were treated with androgens when they were frequent have external genitalia that were masculinized and they exhibited male like sexual behaviour.

Endocrine Environment Behaviour Interactions

Some activational effects involve complex interactions among behaviours, hormones and specific environmental estimate. Will discuss in detail-one example reproductive sequences in ring dives which illustrate this interrelationships.

Reproductive Sequences in Ring Doves

At each stage on this sequence, the internal state of each kind interacts with external variables to produce the observed behaviour pattern. The variables consists of:

1. Environmental eves, such as nests and eggs that influences hormonal and behavioural changes in both.
2. The behaviour of each member of the pair that stimulates changes in the hormonal levels and behaviour of its mate.
3. The hormonal state of both the male female dove, excluding feedback lops.

To delimine whether the presence of a mate or the nesting materials affects incubation behaviours of female ring doves. *Daniel Lehrenan* and his colleagues used three experimental groups.

1. Females housed with a male and nesting material.
2. Females housed with a mate only.
3. Control for males housed alone.

They assessed the results of these pairings in terms of the percentages of test females in each group that exhibited incubation behaviour when presented with a nest containing eggs. Control females never incubated eggs. By days 6,7 and 8 after pairing increasing percentage of females caged with both a male and nesting material incubated test eggs. The concluded that presence of both a female and nesting materials is necessary for complete incubation behaviour in a male.

Now a question may arise that why has finely tuned system of complex interaction among behaviour, hormones and external colours evolved in ring doves. The answer to this question is that since reproductive cycle involves the dual partical ration of both and nececlatis that their activities be coordinated throughout the reproductive success of the pain is guaranteed only if both perform certain acts in syutony e.g. Females that laid eggs before a nest was completed would contributed little to future generations. If the male developed a crop or began to produce crop milk before any rquats had happened etc.

Migratory Behaviour in Birds

Migration in behaviour in birds is also due to the result of the interaction between the environments and the hormones. *Reowan* observed that this peak of gonadal and the recurdiocence coincides with the peak of migration urge. Rowan's suggested that many factors influences migration. Increasing day length increasing ambicint temperature other environmental changes which affect the animal through pituitary activity and perhaps the nervous system directly.

Mechanisms and Ontogeny of Hormonal Effects

Stimuli and Mechanisms

To understand the actions of a hormone, both the stimulation and its release and its effects upon behaviour must be known. The mechanisms by which hormones effects behaviour through their effects upon.

1. The whole organism (e.g. general activity level).
2. Morphologic structure employed in specific response patterns.
3. Peripheral receptor mechanisms.
4. Integrative functions of the C.N.S.
   (a) Control of this development of nervous organisation.
   (b) Control periodic growth and regression of nervous elements.
   (c) Control of sensitivity of stimulation.

Ontogenetic Development

If baby which is injected with testosterone, complex mating responses including covering treading have been observed. There are many examples which ultimately conclude that some interaction of hormonal action and behavioural development establishes a behavioural organization which then persists in the absence of hormonal influences.

Some Recent Developments

Since 1950 research has centered increasingly upon the problem of untangling the delicate and complex interrelationships between hormones, the nervous system and behaviour.

1. In some what similar program of research *Hinde* (e.g. 1985) has been exploring the hormonal bases and external stimulus conditions, controlling nest building.
2. Active research continues to the field of migration too. The *Wolfsen theory* stated the accumulation of fat sets off migratory behaviour has been challenged by the schools.

3. Another interesting approach to hormones and behaviour has been to emplified by the work of *R.P. Michale* regarding sexuality in the verge of *nymphonania*.
4. *D.S. Lehrman* (e.g. 1964) has recently made three discoveries the role of prolactin in parental behaviour in doves.

   First prolactin alone does not elicit parental behaviour, but requires estrogenic *priming*.

   Second low or hormone can act directly on a peripheral structure involved in behaviour. Without involving vagus effects on the C.N.S.

   Third The young birds provide the tactile stimuli that elicit regurgitation from the crop.

Farmer and *King* have shown by ingenious experiments that if captive birds were not allowed to accumulate far *Zugunrube* appears indepently to response to long photoperiod.

2

FEEDING BEHAVIOUR

Feeding is an important functional system because it provides the source of the energy which is used by animals. The systems includes brain mechanisms, with their many behavioural consequences, which result in food intake, and physiological components which involve, firstly, digestion and secondly, the production, storage and translocation of energy-rich compounds. All of these components, together with the nutrient requirements of the body, must be considered in order to understand the allocation of energy to feeding as a functional system and to each energy-requiring process within the system. The description of anatomical and behavioural adaptations for feeding has been a major activity of biologists for many years. Such descriptions, for a wide variety of animals, are available in scientific textbooks such as *Fretter* and *Graham* or *Welty* as well as in many general natural history books.

Psychologists have used food as a reinforcer in a high proportion of experimental studies on learning. Such work has provided much information about the timing of feeding, the quantities of food taken and the amount of work which will be performed for artificial diet rewards in a laboratory. Information about the brain mechanisms controlling feeding behaviour has come from studies of the effects of the stimulation, ablation or chemical treatment of the brain. Other areas of research on feeding have included studies of human nutrition, the relationships between food intake and production in domestic animals, and the feeding behaviour of insect and rodent pests. Recent ideas in quantitative and behavioural ecology are encouraging interactions between research workers with these different approaches to the study of feeding functional system.

Motivational Stage of Feeding

An individual whose motivational state is such that behaviour will be directed towards feeding must take a series of decisions before and during feeding. This series of decisions and the set of mechanisms which makes them possible, is the subject of this chapter. The interrelations between feeding and other functional systems, especially predator avoidance. After a brief discussion of dietary differences and optimality the sections in the chapter refer, in temporal order, to these decisions. Where the word *foraging* is used it refers to the behaviour of animals when they are moving around in such a way that they are likely to encounter and acquire food for themselves or their offspring. Thus, it includes sandpipers finding crustaceans on a beach, sheep grazing in a field, hyaenas hunting wildebeest, and cabbage white butterflies finding and ovipositing on cabbages. Decisions during foraging will depend on potential energy and nutrient returns and on costs. These costs will include those of searching, pursuit, handling and eating; those associated with digestion; those of detoxifying poisonous substances in the food; and deleterious effects of substances in the food.

As *Altmann* and *Wagner* put it 'given the foods that are available to an animal, how much of each should it consume so as to be above its minimum for each nutrient, below the maximum for every toxin, and concurrently, either to minimise some cost function, such as time expended in foraging or exposure to predators while feeding, or else to maximise some benefit obtained from feeding such as protein or caloric intake. Foraging will be affected by previous experience in similar situations so the effects of experience are emphasised in relation to many of the examples quoted. The ecological approach to the study of feeding behaviour is complemented by consideration of physiological mechanisms. These are mentioned in several sections and the section on the control of food intake deals with receptors and brain mechanisms. Most of the studies mentioned in the early sections of this chapter refer to only one aspect of feeding, so two examples of species in which most aspects of the feeding system have been studied, the hummingbird feeding on nectar and the cow grazing on grass are also discussed.

Neural Control of Feeding

Hypothalamic Lesions

The hypothalamus, lying in the floor of the third ventricle of the forebrain of mammals, has important functions regulating the coordination of feeding and many other types of behaviour. Damage to the hypothalamus in rats, cats, mice, monkeys, or dogs may have quite different effects on feeding behaviour, depending on the location of the lesion. A rat with a lesion in the region of the ventromedial nucleus of the hypothalamus overeats until it becomes excessively obese; a 250-gram rat may consume food until it weighs more than 600 grams. On the other hand, a rat with damage or anaesthesia localized in the lateral portions of the hypothalamus will stop feeding and, unless it is kept alive by stomach tube feeding, will starve to death.

Electrical stimulation of these areas has the opposite effect—medial stimulation inhibits feeding and lateral stimulation elicits it. Thus, there seem to be separate mechanisms for starting and terminating a bout of feeding. The onset of feeding is associated with activity of the *feeding centres* in the lateral hypothalamus. This activity involves not only the acts of eating and increased responsiveness to food stimuli but also the appetitive phase of locomotion; lesions in the lateral area result in both cessation of feeding and reduction of locomotion. Eating results in consequences which include activation of the 'satiety centres' in the medial hypothalamus. These centres inhibit the feeding centre, and the bout of feeding behaviour is terminated. Later the inhibition is relaxed, the feeding center becomes, active, and another feeding bout ensues.

The reciprocal relationship between the feeding and satiety centres has been confirmed by an elegant experiment using intracranial self-stimulation. In a series of studies made by *Olds*, rats repeatedly stimulated themselves by pressing a lever in a Skinner box. With electrodes in the first position, the area of the lateral hypothalamus including the feeding centres, stimulation was rewarding in a learning situation. With electrodes in the second position, the area of the medial hypothalamus including the satiety centres, stimulation was either neutral or punishing. By preparing rats with several electrode cannulas permitting both electrical stimulation and administration of a local anaesthetic, *Hoebel* and *Teitelbaum* 91962) were able to study directly the effects of one system on the other. They concluded that the medial and lateral centres control self-stimulation in a manner analogous to their control of feeding.

Anaesthetization of the satiety centres resulted both in more feeding and in more self-stimulation in the feeding areas. Stimulation of the medial satiety centre inhibited feeding and also inhibited self-stimulation in the feeding areas. Stimulation of the medial satiety center inhibited feeding and also inhibited self-stimulation of the lateral feeding centres.

Previous feeding to satiation also inhibited lateral self-stimulation for periods from 30 minutes to 2 hours. A hungary animal engaged in more lateral self-stimulation, presumably with minimal activity of the satiety centers. Although the hypothalamus clearly plays a major role in regulating food intake in mammals, it is not the only brain centre concerned with this function in mammals. Lesions in the amygdala may also result in overeating in cats, although the hyperphagia is only one-third of that caused by medial hypothalamic lesions in the same species. Other areas of the brain may be more concerned with discrimination and a predilection for certain food items.

Onset of Feeding

We know from behavioural evidence that the activity of centers controlling the facilitation and inhibition of feeding must be related to the nutritive state of the animal. There is still doubt about how this balance is maintained. Consider the onset of a feeding bout that begins

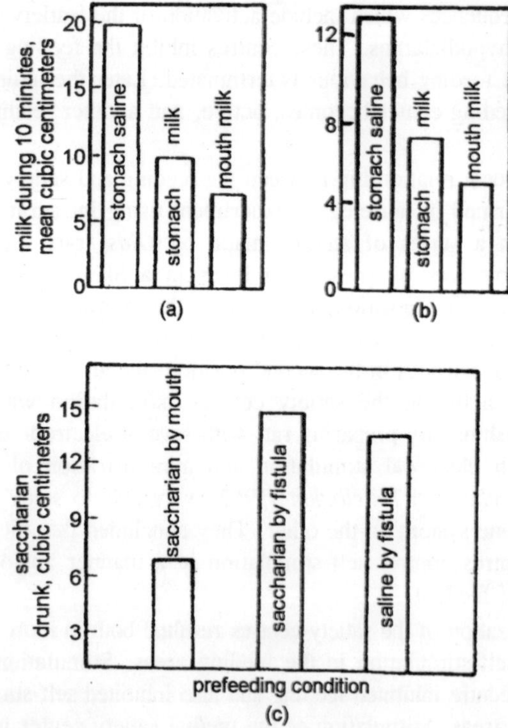

Fig. 2.1. The amount of milk drunk by a rat in a 10-minute test period is affected by previous treatments.

with appetitive behaviour, rather than one that is immediately elicited by external stimuli from food. What activates the feeding center and leads to the onset of locomotion which results in feeding? Early worker thought that stomach contractions, which occur after a period without food, were instrumental in starting feeding behaviour. After long periods of food deprivation they may have an important influence. But various lines of evidence suggest that their rate of contraction is unimportant in normal feeding. A man with a denervated stomach still eats regularly.

Moreover, studies of the effects of various stomach conditions on electrical activity in the hypothalamic centers have shown that stomach contractions are not correlated with activity in the feeding centers, though they are inhibited during activity of the satiety centers. The evidence indicates that some parts of the brain, perhaps the feeding and satiety centers, respond directly to properties of the circulating blood and so instigate feeding behaviour. Alternative or complementary suggestions about the eliciting stimuli, often well supported by evidence, include: a small difference in concentrations of glucose and other metabolites in the arterial and venous circulations, slight drops in the body, and fluctuations in the concentration of a hormone. *Brobeck* concludes that several of these factors and perhaps all will prove to be involved, but that it is not possible at the present time to assess quantitatively the relative significance of each.

Duration of Feeding

Once the hungry animal has located food, feeding behaviour is elicited by the external stimuli provided by the food. What factors determine how long feeding will continue before the bout is terminated? Various consequences of feeding are involved. In the first place, the repeated eating and swallowing may itself eventually be self-exhausting as a result either of performance of the motor activity, or of stimulation of receptors in the head and throat. Thus an animal with a fistula which discharges food from the esophagus to the outside rather than to the stomach eats much more than usual. But it does not eat continuously; there are still bouts of feeding. A given quantity of food eaten through the mouth is ore effective in reducing subsequent feeding than the same quantity of food given directly into the stomach.

As a corollary, it has been established that an animal which is stomach-fed through a fistula with as much or more food than it normally eats in a meal still shows an appreciable amount of feeding behaviour. It is therefore clear that oropharyngeal factors participate in terminating about of feeding behaviour, though seldom with a dominant role. Arrival of food in the stomach is critical in the process of satiation. Distension of the stomach wall stimulates stretch receptors which activate the satiety centers in the hypothalamus. Partial inhibition of further feeding can be induced by filling the stomach with inert material. Food is more effective however. Miller has shown this by comparing the effects on subsequent drinking of enriched milk, of enjecting similar quantities of milk or isotonic saline directly into the stomach. Furthermore, an animal given food that is diluted by varying amounts, of inert material will compensate to some extent by eating more. The amount of stomach distension permitted before feeding stops must somehow be related to the nutritive properties of the food; chemoreceptors in the stomach may be responsible. Some further consequences of eating may come rapidly into play, such as the flow of fluids into the digestive tracts accompanying eating, the consequent dehydration of certain tissues, and the release of acetylcholine and other chemical mediators during intestinal absorption.

In general, the assimilation of food is too slow to play a direct role in terminating feeding. Several investigators have pointed out the lack of relationship between the nutritive value of food consumed in a given bout of feeding behaviour and the termination of that bout. A longer-term relationship necessarily exists, resulting in an overall correlation between the nutritive state of the animal and food intake. In this general maintenance of homeostasis, such factors as levels of glucose and other metabolites in the blood presumably come into play .The rise in temperature consequent upon assimilation of a meal may also play a part. In a number of animals warmth and hyperthermia inhibit eating while cold environments encourage it.

Feeding in Relation to Body Weight

A possible relationship between body weight and food intake is suggested by Teitelbaum's studies of hyperphagic rats. The specific effects of lesions in the satiety centers in causing overeating have already been discussed. In one of the few studies of the detailed temporal patterning of feeding behaviour which have been published, he found that hyperphagic rats given an enriched fluid diet eat larger meals, but that they drink at a normal rate and eat meals with the usual frequency with a solid diet, the increased food consumption in hyperphagia is associated with more frequent and larger meals. Teitelbaum associated this difference with the greater caloric value of the liquid diet. He went on to show that dilution of the liquid diet induced a predicted increase in the frequency of meals in both hyperphagic and normal rats.

Evidently the hyperphagic rat, though it overeats, still regulates the caloric intake of food. Lesions in the satiation centers do not cause completely unregulated feeding behaviour, but something more subtle. Teitelbaum notes that behaviour of a hyperphagic rat changes strikingly when it reaches a certain degree of obesity, and its feeding behaviour comes to resemble that of a normal rat. An obese hyperphagic rate ceases to overeat and consumes only enough to maintain its overweight condition in a steady state. This behaviour suggests that body weight or some correlate of it is another factor related to the organization of feeding behaviour and that this relationship os disrupted by lesions in the satiety centres. As Teitelbaum puts it, "one might say that the hyperphagic animal overeats to get fat. Once it is fat, it no longer overeats." This notion is reinforced by demonstration that in rats which have been made excessively overweight by other means, lesions cause no hyperphagia. Teitelbaum concludes, with *Kennedy* that the satiety centers may be directly or indirectly responsive to some circulating metabolite related to the state of fat deposits in the animal's body.

General pattern of Feeding Mechanisms in Mammals

The tentative picture drawn from these studies of the physiological mechanisms underlying feeding behaviour is as follows. A bout of feeding behaviour starts with locomotion in search of food— the appetitive phase. This phase is triggered by activity in the lateral hypothalamic feeding centers and inactivity in the medial satiety centers. The release of the feeding centers from inhibition by the satiety centers, a response to a wide variety of changes on the body associated with deprivation of food, may be the primary step. Changes in higher brain centers may also release feeding areas from inhibition, allowing for the demonstrable influence of learning on periodic patterns of feeding. After repeated acts of eating and swallowing, followed by arrival of food in the stomach—the consummatory phase—the bout of feeding eventually terminates.

Oropharyngeal factors play an undetermined but definite role. Stomach distension plays a critical role, directly activating the satiety center, which then inhibits the feeding center. Passage of fluids into the gut may have a similar effect. The maximum stomach distension permitted varies with the nature of the food material so that diluted food is taken in greater quantities. The general nutritive state of the animal may also determine, within limits, the degree of stomach distension permitted and thus the duration of the feeding bout. Perhaps the nutritive state directly affects the threshold of responsiveness of the satiety center.

The animal then refrains from feeding for a time, even though it may still be exposed to external stimuli from food materials. The length of the interval before another bout of feeding begins seems to depend on a wide variety of factors including temperature, body weight, and concentration of several metabolites in the blood. In general, it appears that the greater the nutritive deficit, the more rapidly the activity of the satiety center wanes. Feeding resumes more rapidly, and more food is consumed in each bout. Although the duration of feeding bouts and of intervals between bouts varies with the nutritive state of the animal, the rate of actual feeding movements seems to very relatively little under normal conditions.

We are now in a better position to understand the apparent inconsistencies by Miller and *Tugendhat* in the changes of motivation during cycles of feeding and satiation. There is not single, unitary, underlying mechanisms but a multiplicity of mechanisms which are at least in some degree independent of one another. The contrasting results obtained by applying different measures to the same sequence of feeding are no longer paradoxical if we know that many or all of these various physiological factors can influence the animal's behaviour with relative degrees of independence. The sum of their effects is the rhythmical pattern of bouts of feeding separated by periods of other kinds of activity.

This repetitive cycle has no simple basis in exogenous or endogenous influences. The underlying mechanisms involves a subtle integration of effects of stimuli impinging on organisms from without and changes taking place within the organism. External influences do of course loom large, for the behaviour is designed as part of a homeostatic mechanism which must make allowances for variations in the availability, palatability, and nutritive value of different types of food. It is appropriate to consider their contribution in more detail.

Nature of Diets

Feeding behaviour is often dividend into separate categories according to whether the food material comes from living animals, living plants or dead organisms. Animal morphology is, to a large extent, a result of adaptations for feeding so there are considerable differences between some animal-feeders and some plant feeders, but the above categorisation of feeding behaviour is not always useful. Animals which are suspension feeders may accept animals, plants or organic detritus as food. A warbler may hunt for berries or for insect larvae by methods with similar components. Limpets grazing on rocks eat algal sporelings and barnacle larvae, whilst a nudibranch molluse which is grazing on hydroid colonies will show behavioural similarity in its feeding methods to an insect or mammal grazing on plants. Different digestive mechanisms are required for some components of animal and food but there is much variation within vegetable and animal diets and some plant eaters actually absorb the products of commensal animals (Protozoa) or bacteria.

Although there are of the similarities in the feeding behaviours, digestive processes, etc., of animals eating food of diverse organs, some aspects of feeding are specific to one type of diet. The digestive process in grass-eating ruminants has no parallel amongst species in which the diet is composed of animal materia. Likewise, no plant-eaters needs to chase prey actively so none has need of the adaptations for efficiency of capture possessed by a wolf or a peregrine. Some herbivores are very different in their feeding from some predator but the terms are less useful when considering the many animals with intermediate feeding behaviour and physiology. The most important behaviour which makes feeding possible for parasites in usually host seeking. The behaviour of a parasite looking for a host should perhaps be considered as a form of habitat selection but some of the behavioural mechanisms are similar to those shown by non-parasitic animals when hunting for food.

Arm Races

Daskins and Krebs have pointed out that the evolution of predator and prey species can be regarded as an arms race. Modifications of genotype continually result in an improvement in the chance that the average individual fox will catch a rabbit or that the average rabbit will be able to evade foxes. A similar arms race exists between herbivores and plants, for plants can produce toxic or unpleasant tasking substances and can bear spines, hairs and other deterrents to would be consumers. Animals can modify their digestive systems and feeding methods to avoid the adverse effects of there plant weapons, Plants and their ability to modify their growth form and their ability to grow in places which are inaccessible to the animals. This arms race is a very slow one which takes many generation and the characteristics of a potential consumer, or a potential food, represent only a few of the many factors which might influence evolutionary changes. The evolutionary competition between lettuces and rabbits is, however, just as real as that between rabbits and foxes. Animals which eat dead matter compete with other organisms which are trying to consume their food, rather than with the food itself.

Feeding Categories

A major factor which effects feeding behaviour is the distribution of food. Distribution must be assessed in relation to the size and locomotor ability of the animal. Some animals are surrounded by readily accessible food whereas others may live for long periods without encountering a food item. Before reviewing behavioural methods for finding and acquiring food it will be useful to consider the degree of dietary specialisation or generalisation which is shown by animals. The extent of specialisation which is possible depends upon the abundance and the spatial and temporal distribution of possible food items. Most specialists eat food which is locally or generally abundant but generalists can eat foods which they encounter rarely. Animals which eat only one type of food are called *monophagous*. Those which eat a limited range of foods are referred to as *oligophagous* whilst those which will eat a wide variety of foods are *polyphagous*. Examples of species which are usually monophagous include the giant panda (*Ailuropoda melanoleuca*) which eats bamboo shoots, the snail kite (*Rostramus sociabilits*) which eats the freshwater snail *Pomacea*, the grasshopper (*Gesonula punctifrons*) which eats the water hyacinth (*Eichhornia crassipes*), the chalk-hill blue butterfly caterpillar (*Lysandra coridon*) which eats the leaves of the horse-shoe vetch (*Hippocrepis comosa*), and many parasitic species. Animals which are oligophagous include the koala (*Phascolarctos cinereus*) which eats leaves from about five species of *Eucalyptus* trees, the brent goose (or brant) (*Branta bernicla*) which seldom eats anything other than species of eel grass (*Zostera*), the grasshopper (*Chorthippus parallelus*) which eats grasses, and the larvae of butterflies in the family Pieridae which eat plants of the family Cruciferae with mustard oil in their leaves. Many animals are polyphagous. Examples of species of similar sizes and from the same groups of animals as those mentioned above are bears, rats, man, the jay (*Garrulus glandifarius*), the great bustard (*Otis tarda*), the locust *Schistocerica gregaria*, and the moth *Spodoptera littoralis*.

It seems likely that the ancestral forms of most animal groups were polyphagous. *Emlen* points out that if the members of a species encountered one food much more often and therefore ate it more frequently than others which provided the same benefit in terms of fitness, the average ability to find, ingest and digest that food would probably improve. Such dietary selection would, in a few generations, result in this food being of greater benefit than other foods. Genes which promoted a preference for this food would then spread in the population and the first step towards monophagy would have been taken. The digestion of one food requires a smaller range of enzymes than is required by a polyphagous animal. The anatomy and the behavioural repertoire can become more specialised and hence more efficient if only one food is eaten. *Levins* and MacArthur emphasised the monophagy might be advantageous, in terms of the number of young produced, because there is less risk that an inadequate food source would be chosen. A major problem for monophagous animals is what to do when the food source is scarce. Scarcity is a relative term here for it refers to food which cannot be found in quantities adequate for survival by individuals of that species. Some species live through periods during which the food upon which they specialise is absent, by accumulating fat reserves, hibernating, aestivating, encysting, producing long-lived eggs or by hoarding the food when it is abundant. Food is hoarded by squirrels, rats, acorn woodpeckers (*Melanerpes formicivorus*), ants and man. Specialist feeders thus behave in a way which minimises the variance in the availability of their food when they require it. Animals living in unstable habitats cannot to be specialists and thus require a mor complex repertoire of feeding behaviour.

Polyphagous animals can exploit food sources which are available briefly and they risk less if they move away from a food source because of predator activity or competition. Monophagous animals can often be found easily by predators which merely search for the food source but polyphagous animals are less readily found in this way. Another advantage for some polyphagous animals is that they do not have to range so far in order to find food so they can live in a home area for long periods. Both polyphagy and monophagy have advantages and the effectiveness of one as a strategy in a habitat will depend upon the diversity and stability of population of potential food organisms and on the strategies adopted by competitors in the habitat.

Optimality Approach to Studies of Feeding

An assumption which is made by many of those who speculate upon the mechanisms of feeding behaviour is that, since the genotype is the result of a long period of evolution by natural selection, the behaviour of individuals today may be close to the optimum for the circumstances. A consequence of this assumption is that the energetic costs and benefits of feeding behaviour have been measured in many studies. The result has been a great improvement in our understanding of feeding mechanisms. The optimal foraging approach advocated by MacArthur and *Pianka* and Emlen has been vindicated by combined studies of feeding ecology, physiology and behaviour.

Any attempt at cost-benefit analysis of feeding must take account of physical constraints, digestion costs and the necessity for acquiring the complete range of essential nutrients as well as the energetic costs of feeding behaviour and the energetic returns from the food. However, it is not sufficient to access optimality in terms of energy obtained in relation to energy expended. The ultimate measure of optimality must refer ideally to the spread of genes in the population. In the absence of altruistic behaviour towards individuals other than offspring, the long-term reproductive potential of the animals under consideration is the factor which is optimised. Statement about optimality of feeding strategies are not useful unless antipredator behaviour, avoidance of other hazards, the regulation of body temperature etc., are also considered. This same argument applies to statements about whether a feeding strategy is evolutionarily stable.

Finding Food

Members of most animal species have a dispersive stage in their life history so that they need to have the ability to distinguish between areas where food might occur and those where it will not. Once such an area is found, other behavioural mechanisms which increase the chance of encountering food items and make possible their acquisition or rejection, come into operation. The area can then be assessed, the effects of food intake monitored, and further behaviour modified feeding behaviour of some animals includes all of the aspects mentioned in whereas that of others is much simpler. Monophagous animals which spend most of their life eating one individual plant or animal may still have to take decisions about which part of the food organism to eat but their feeding behaviour is less complex than is that of a polyphagous animal which can feed in various habitats.

Finding an Area

Patches and habitats

The distribution of the food of most animals is clumped so that the individual must find area os local concentration amidst areas of low concentration or absence. Many people call such an area of local food concentration a patch. Before finding a patch, the individual must find the sort of habitat where sources of its food might occur. The importance of food in determining what can be suitable habitat is emphasised by *Hassell* and *Southwood's* description of a habitat as a collection of patches. For most animals the habitat must have other characteristics as well. Animals often live in a particular area long enough to learn many of its characteristic. Such familiarity is beneficial when evading predators and reducing physical hazards, as well as when seeking food.

Examples of patch-finding include a squirrel finding a nut tree, a mountain sheep finding an area of grass on a rocky mountainside, a goshawk finding flock of pigeons in a field, or a ladybird beetle finding an area on a plant with an aggregation of aphids on it. The amount of food which can be regarded as a patch for a large animals may be a lifetime's supply for a small one. One fruit may sustain a moth caterpillar for the weeks during which its larval development occurs but may be consumed in seconds by a monkey might be, for example, one, several or many ripe fruits on a tree. Species like fruit-eating monkeys often have to travel for sometime between patches whereas leaf-eaters exploit much larger patches and travel less. For example, in a study of the spider monkey (*Ateles geoffroyi*) the animals were shown to spend 28% of daylight in travelling from one forest tree bearing ripe fruit to another. In contrast, the similar-sized black and white colobus monkeys (*Colobus guereza*) ate leaves during 77% of all observed feeding time and travelled for only 5% of the day. These figures are typical of a general relationship for primate species between time spent travelling and proportion of foliage in the diet.

Although there is variation amongst species in the amount of time spent searching for patches, this behaviour is very important for almost all animals. Efficient searching will result in finding better quality patches and hence improve the reproductive potential of the individual.

Habitat selection experiments

Casual observations of habitat selection have been reported frequently and some detailed experimental studies have been carried out. *Wecker* found that two subspecies of the deer-mouse (*Peromyscus maniculatus*) occurred in two habitats, grassland and mixed oak and hickory woodland. Members of the grassland subspecies kept in an enclosure with grassland on one side and woodland on the other showed a clear preference for the grassland: they are, presumably, better adapted for feeding etc., in grassland. When Wecker reared some

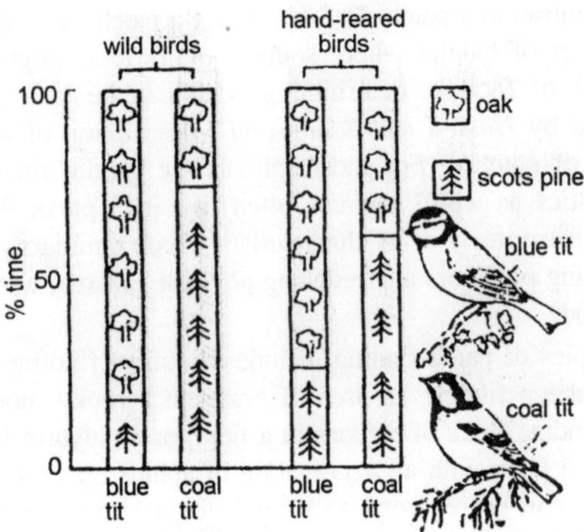

Fig. 2.2. The proportion of time spent in oak or scots pine branches by blue tits and coal tits given a simultaneous choice. (a) Wild birds; (b) Hand-reared birds.

mice in the grassland habitat and others in laboratory cages the individuals reared in grassland showed a stronger preference for it. If the mice were reared in the woodland habitat, they also showed a preference, albert weaker, for the grassland habitat.

Observations by *Gibb* showed that coal tits (*Parus ater*), which feed most frequently in coniferous trees, preferred coniferous branches in laboratory experiments whereas blue tits (*Parus caeruleus*), which feed in deciduous trees, preferred deciduous branches. Partridge reared tits of both species in a laboratory environment where they had no experience of any vegetation. When full-grown, the coal tits showed a preference for pine branches whilst the blue tits preferred oak branches. Although the preference was not as marked as that of birds caught in the wild, Partridge's experiments demonstrates that the mechanism which determines the preference can develop in the absence of any experience of the preferred trees. Partridge's conclusion that the preference is 'genetically determined' could be taken to imply that environmental factors are not involved in the development of the mechanisms controlling the behaviour.

All behaviour depends on genetic and environmental factors but Partridge's results do show that input which results from contact with the trees is not amongst the environmental factors which affect the development of this preference. Further experiments and observations of the skills associated with feeding in coniferous or deciduous trees showed that the tits preferred the trees for which their skills were best adapted. Studies on a variety of species show that some aspects of habitat preference are modified by experience whilst other are not.

Host-finding behaviour

Suitable habitats, and consequently food sources, can be found by moving around and responding to visual and olfactory cues; sometimes at a considerable distance from the habitat. The birds and mammals mentioned above respond to a complex set of sensory cues but an Anopheles mosquito which emerges as an adult after its aquatic early life can find human dwellings by a comparatively simple strategy. If flies across wind until it encounters the odour from human dwellings and then flies up-wing. Such behaviour is similar to that of active aquatic predators, such as sharks, which will swim up concentration gradients towards bodies which are emitting blood and the starfish (*Asterias*) which will move up-stream readily when the body will fluids of their molluscan prey are present in the water.

Animals which are not sufficiently mobile to find the required food source have to modify their own behaviour, or modify their surroundings, in such a way as to maximise the chance that they will encounter their food. The habitat of the sheep tick (Ixodes ricinus) is

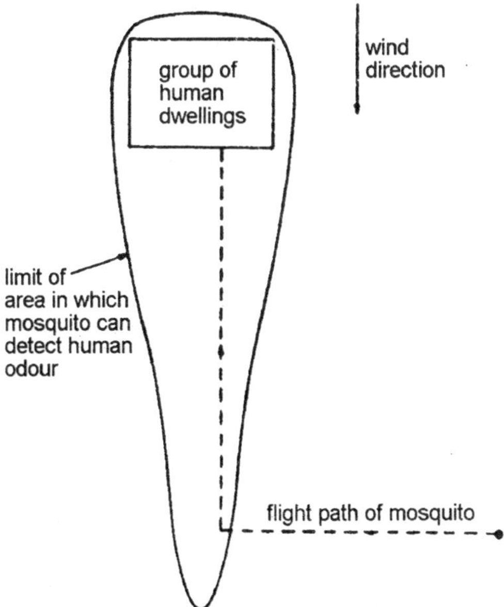

Fig. 2.3. The mosquito Anopheles gambiae finds human-dwellings by flying across wind and then upwind when odour is detected.

a pasture with sheep in it and within this habitat, a pitch is a sheep. Sheep ticks increase the chance of encountering a sheep which comes near them vegetation. Provided that they do not lose too much water they remain at the tops of the stems until they are disturbed by something whose characteristics are, ideally for the tick, that it is moving, warm and woolly.

Internal parasites are extreme examples of animals which have problems in finding the next feeding site. There is a large literature on the complexity of parasite life cycles but amongst the most interesting species are those which hi-jack one host or just hitch-hike in order to reach the next. Many parasites have a general debilitatory effect on one host which is then more vulnerable to attack by the parasite's next host. Some parasites, however, modify their environment, that is to say the host, in such a way that the host's behaviour is changed. The acanthocephalan parasite *Polymorphus paradoxus* lives in freshwater amphipod crustacean *Gammarus* and has to get to it next host, the mallard duck (*Anas platyrhynochos*). *Bethel* and Holmes found that, whereas 97% of uninfected *Gammarus* were found in dark areas in an experimental apparatus, only 29% of infected *Gammarus* remained in the dark. The behaviour of the other 80% was altered by the presence of the parasite so that they swam in the light and clung to objects near the water surface. Here, they are much more likely to be eaten by ducks than are individuals which remain in the dark areas near the bottom of a pound or steam. A study of the effects of the eyefluke *Diplostomum spathaceum* on the behaviour of fish showed that heavily parasitised dace and trout spent more time swimming near the surface of the water than did those with few parasites. Here they are more likely to be eaten by gulls which are the nest host of the flukes.

Finding patches by observing conspecifics

The importance of responding to other individuals, often in the same social group, which move in directions which might lead to a food patch is already described. These ideas were initiated by observations of small birds joining flocks which move towards patches, but there are also examples of large birds in the air keeping visual contact with one another so that when one descends, although it might prefer not to share its find, the others fly towards it. This patch-finding technique is used by vultures finding carcases and by seabirds finding concentrations of fish near the surface of the sea. Even if the individuals moving towards a patch are not seen, the patch itself may be rendered much more conspicuous by the presence of an aggregation of conspecifics or individuals of other species which may feed in similar patches. The presence of these individuals is an indicator of environmental quality. The presence of these individuals is an indicator of environmental quality. In experimental studies, Krebs found that great blue herons (*Ardea herodias*) were more likely to land at possible feeding sties at which he had put models of herons.

Finding Food Items

Aggregation in patches

The efficiency with which patch-finding mechanisms operate in a variety of species is attested to by the aggregation of animals in regions of high food density. Such aggregation occurs in species in which individuals find their way to patch separately as well as in species whose members normally move around in groups. Some examples of animals whose distribution is much affected by that of patches of food. The occurrence of the aggregation is due in part to the use of similar patch-finding mechanisms by many individuals but is accentuated by the greater likelihood that individuals will stay in an area of high food abundance. The advantages of staying in the patch may be counteracted, however, by the disadvantages which result from competition with the other individuals in the patch.

Scanning, recognition and reactive distance

Many animals enhance their chances of finding food by scanning their surroundings. A monkey looking for a fruit, a mantis looking for a fly, or an owl sitting on a perch at night listening for the rustle caused by a mouse, all adjust the head position so that the receptors can detect and locate the food. Such scanning, whether or not accompanied by locomotion, may account for a high proportion of foraging time, for example, in the thrushes observed by Smith. It may be used during patch finding as well as during the detection of items within a patch and is an important factor in food finding by many species.

The *recognition* of food items depends upon the functioning of sense organs, together with sensory analyses in the brain and on the operation of attentional mechanisms. The efficiency with which sensory analysers function is a factor which limits of recognition will also depend upon the background against which the item must be detected. Most research on this topic refers to detection by means of vision or olfaction but some animals use other senses. They may have to modify the environment in some way, e.g., by digging into the ground or by moving vegetation, before they encounter a food item.

Precise evidence about the distances at which food is perceived by different species is sparse although we as individuals know the approximate distance at which were recognise mushrooms, berries, shellfish or other items of which we are hunting. Estimates of *reactive distance* have sometimes been made in studies of feeding. *Beukema* recorded the distance at which sticklebacks (*Gasterosteus aculeatus*) turned towards or accelerated towards food items. When the food was the small red worm *Tubifex* the reactive distance was 25 cm. For another small fish, the dace (*Leuciscus leuciscus*) the reactive distance to the amphipod crustacean *Gammarus* varied from 20 to 50 cm according to the numbers of eyeflukes (*Diplostomum*) in the eyes of the fish.

The effect of experience on the reactive distance is considerable in situations where the hunger does not know, initially, the precise characteristics of the prey. When Beukema started to provide sticklebacks with the larvae of the small fly *Drosophila*, they failed to react to the prey at 10 cm distance on most of the first twenty presentations. By the fifth set of twenty presentations, however, they reacted to almost all larvae at that distance and to some at 20 cm. No stickleback reacted to *Drosophila* larvae at 30 cm but some did react at this distance to the larger, red, wriggling *Tubifex*. The reactive distance to a food item, after extensive experience of feeding on that food alone, is likely to be limited by sensory ability but attentional mechanisms are more likely to determine receive distances in other situations.

Search Paths

The movements of individual animals searching for food within a patch have been studied in detail for several species of birds and insects. The search is unlikely to be random for this would involve repeated search in some area and would be inefficient unless the food resource was renewed very rapidly. Ornithologists have long known that owls systematically 'quarter' fields when hunting for small mammals and that feeding flocks of pigeons or starlings move around fields in such a way that they do not recross their own paths. The direction of movement of an owl is, approximately, a straight line except at the edge of the field where the owl swings around so as to return across the field on a path parallel to the first. The distance between successive traverses of a field must depend upon the distance from which the food item can be recognised and attained by the owl. The area around a prey species which is a circle whose radius in this distance, has been called the *danger zone*. A similar term, but one which takes into account the presence of other prey individuals, is the *domain of danger*. These observations of owls and other hunters also indicate that predator species control their movements precisely, use short-term memory of the topography when computing courses and make decisions about what constitutes a boundary of the hunting area. Taking into account all of the factors mentioned above it is possible to formulate models of the optimal searching paths of different hunters in different situations and several authors have attempted to do this.

Detailed descriptions of searching paths were made by Smith who watched blackbirds (*Turdus merula*) and song thrushes (*T. philomelos*) hunting for earthworms in a meadow. The birds moved forward for about 0.5 s and then paused to scan for a mean of 4.8 s before moving again. The mean move length was 34 mm for male blackbird and 450 mm for song thrushes. If no worm was found, the succession of moves often included alternate right and left turns and the beeline direction was approximately straight. If food was found, the thrush often made two or more turns in the same direction and the beeline distance of the twelve moves after capture was shorter than that during the ten moves before capture. In some experiments artificial food, in the form of pastry caterpillars, was provided. Smith found that the search path included more turns and less alternation of turns when thrushes, which had been moving through an area of low food density, encountered an area of high food density. He called this behaviour area concentrated search, a more precise description than 'area restricted search' which other authors have used.

Drent, *Tinbergen* and *Tinbergen* and *Drent* obtained a similar result from observation of starlings hunting leatherjackets, larvae of crane-flies in the family Tipulidae, in grassland. When starlings were hunting for food to take to their nestlings, they were observed to search in comparatively straight lines until leatherjackets were found, then to show area concentrated search. When they had obtained a supply of

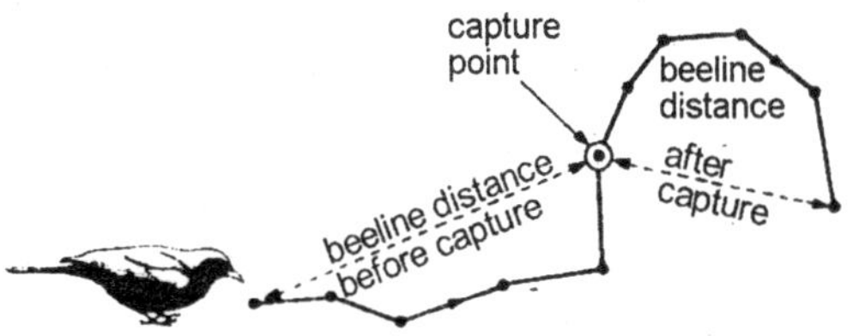

Fig. 2.4. Movements of a blackbird hunting for worms. The beeline distance during five moves before capture is longer than that after capture.

food they flew off to their nest and then returned to the spot where the high concentration of food had been found. Not only do the birds remember the previous rate at which food had been encountered, when deciding whether or not to adopt area concentrated searching, but they remember localities in open grassland with great precision.

The search paths shown by captive oven-birds (*Seiurus aurocapillus*) hunting for mealworms. *Tenebrio* bettle larvae, in a 6m × 6m grassy arena were also affected by present and previous food density. When most of the area of the arena was subdivided into four paths which were provisioned with mealworms at densities of 0.9, 1.8, 3.6 or 7.1m^2, the number of visits to a path and length of search path in a patch were much higher in the highest density patch. The logarithms of the number of visits and of the search-path length were proportional to food density. The length of path per visit and the meander ratio (actual path length/beeline distance) were proportional to food density. The overall result of this searching behaviour was that the highest proportion of prey was taken from the highest density patches. When over-birds which had fed in an arena provisioned in this way were returned to it on the following day they immediately started searching in the area which had been most profitable on the previous day. If there was no food there they soon search elsewhere and if there had been two profitable patches they visited both. *Zach* and Falls also provide evidence for the avoidance by ovenbirds of areas which have been searched thoroughly, whilst Thomas describes the avoidance by sticklebacks of areas in which possible food items had been found but rejected.

Complex search paths are also found amongst invertebrates, for example the bumble bees *Bombus* studied by *Pyke*. The bees collect nectar and fly from flower to flower in a field on comparatively straight paths but with alternation of left and right turns. In accounting for all aspects of the search paths shown, Pyke concluded that the bumble bee must remember its arrival direction at a flower, its change of direction at a flower, its change of direction t the previous inflorescence and the amount of nectar obtained from the flower just visited.

Foraging

Item size

The ways in which animals deal with situations where there is only one type of food item have been considered. Alternative food items may vary in size, and hence in energy return when the item is ingested. Just as there are components in feeding behaviour which increase the probability of obtaining more food items per unit of energy expended, optimal foraging theory predicts that if items give a greater net energy return they are more likely to be selected.

Where animals of various species are offered the opportunity to search for and to acquire food items of different sizes, such that the larger items are more energetically rewarding, the proportion of larger items taken is larger than that available in the population, especially at high food densities. Examples include bluegill sunfish (*Lepomis macrochirus*) eating the crustacean Daphnia great tits eating mealworm pieces redshank (*Tinga totanus*) eating Nereis worms and shore crabs (*Carcinus maenas*) eating mussels (*Mytilus edulis*).

Nutrient quality

If the potential food items which might be encountered differ in quality as well as in size, animals use behavioural mechanisms to maximise the chance that they will ingest items which provide the most energetic and nutrient benefit. In polyphagous species a mixture of different types of food is often needed. The mechanisms used to obtain the food items of the best size and quality often involve taking decisions about searching behaviour and about whether or not to acquire an item once it is located. Sometimes changes in searching behaviour are obviously dependent on previous experience. The starlings studied by Tinbergen and Drent (1980) fed their young on a mixture of foods of which the majority (55-80%) were leatherjackets but there was always a proportion of caterpillars of the moth *Cerapterix*, even when leatherjackets were plentiful. A starling which had been collecting leatherjackets from grassland and bringing them back to its nest would fly off in a different direction to a specific area of the salt-marsh where the *Cerapterix* larvae where to be found.

Other examples of observed changes in food searching come from primate studies, for example *Chivers* found that the siamang *Symphalangus syndactylus*, an ape which he observed in Malaysia, would climb to the ends of branches in the early part of the day and eat fruit. At some point in the morning it switched to the central trunk region and ate leaves, thus obtaining a variety of nutrients each day. Much more complex sequences of movements to different feeding sites are observed in species of primates which eat a greater variety of foods. Baboons will move from site to site eating different types of flowers, fruits, grasses, rhizomes, insects, birds and mammals. Many of the sites visited by these and other animals are familiar to the individual so it knows where to search as well as how to search for

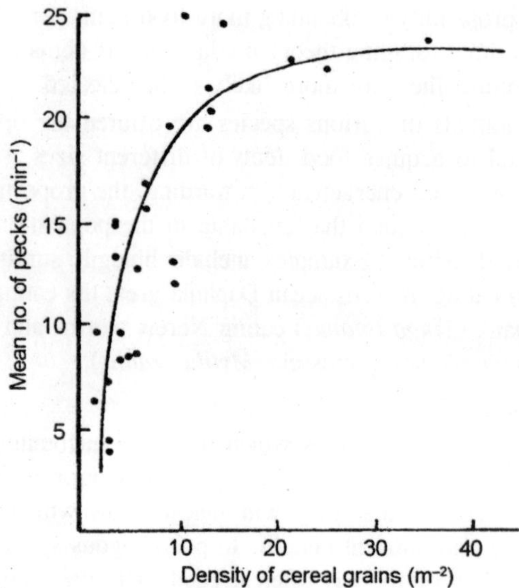

Fig. 2.5. The rate of pecking at cereal grains by woodpigeons feeding in a field has minimum and maximum values as shown.

this type of food. Whilst searching for food in any particular area, however, it is possible that more than one type of food might be encountered. Attempts to relate searching methods to the frequency of occurrence of each of several types of food in a feeding area have been the subject of much research since the studies of L. Tinbergen (1960).

Item density

The relationship between the number of food items acquired by an animal and the density of occurrence of these items were described as type 1, type 2 and type 3 functional responses by *Holling*. In his *type 1 functional response* the rate of acquisition is directly proportional to density up to maximum acquisition rate. He describes the occirrseme of such a response in animals which are filter feeders. The other two types of functional response occur as a result of behavioural mechanisms and are explained in this section. An example of the parabolic *type 2 functional response* whilst the sigmoid *type 3 functional response*. In the example the rate of pecking at newly sown cereal grains by woodpigeons (*Columba palumbus*) varied according to grain density, but no pecking occurred at densities of less than 2m^{-2} and no further increase in pecking rate occurred at densities of over 30m^{2}. *Murton, Isaacson* and *Westwood* suggested that the lower limit was due to the low nutrient return per unit of energy expended at this density whilst the upper limit was a consequence of the time taken to acquire the grain. The importance of this *handling time* was further emphasised by Holling in experiments with human and insect hunters.

Tinbergen recorded the frequency with which small insectivorous birds, especially the great tit (Parus major), brought insects of different species to their young. He related these results to the frequency of occurrence of these insects in the pine-forest habitat of the birds. Most rare insects were taken by the birds less often than chance would lead one to expect and common ones more often. This result is very similar to that obtained by *Holling* who studied feeding by small mammals on sawfly larvae in grassland and in laboratory experiments. Most of the food items, in these two sets of experiments, were cryptically coloured and Tinbergen proposed that they were not recognised as food when the rate at which they were encountered was low. Once the encounter rate rose above a certain level the items were taken by the hunger, which they adopted a *specific search image* for this type of food. This idea of modification in perceptual mechanisms has been widely criticised, for example by *Dawkins* and by *Royama* who reported that tits often did not collect the same type of food items on successive visits to the same area.

Despite the existence of data which do not fit Tinbergen's theory, there is no doubt that individual hunters do sometimes show long sequences of taking one type of food when another energetically equivalent food is readily available. Murton conducted a field experiment with wood pigeons feeding on cereal stubble or clover pasture. He provided maize, tic beans, maple peas or green peas all of which had been treated with a-chloralose, which stupefies pigeons. After sometime he collected the pigeons and examined the contents of their crops. When equal amounts of tic beans or maple peas were evenly scattered in a field, one pigeon ate only tic beans whilst others ate only ample pears, and the majority showed a clear preference for one or the other food. Murton points out, however, that due to social facilitation within the flock, his data do not fit with all the ideas produced by Tinbergen: Discussions about methods of obtaining food are based on a variety of casual factors so a complete understanding of what happens during foraging necessitates consideration of the effects of experience, motivational mechanisms and the attentional mechanisms which they modify. Tinbergen's search-image concept is clearly an oversimplification of hunting strategy in general. If interpreted in terms of attentional rather than sensory mechanisms, however, it helps in the understanding of some aspects of feeding behaviour.

When to give up foraging in an area

If food is not found in an area the animal must have a mechanism for deciding when to give up searching there. Such a mechanism would also be used after some food had been consumed but no further food was being found. The *giving up time*, during which no food is found, was studied by *Krebs*, *Ryan* and *Charnov*. The likelihood of giving up should increase as environmental quality declines. Charnov emphasised that the forager must relate intake in the patch to average intake possible elsewhere. It is also necessary for the forager to consider the costs of getting to another patch and the variability of food supplies. This problems is considered further in the section on hummingbird feeding, by Pyke et al., and by Pyke as well as by the other above mentioned authors.

Eating: Energy, Nutrients and Dangers

Once a potential food item has been found by an individual, the decision as to whether or not to attempt to acquire it may depend upon a variety of factors. The item will provide benefit in terms of energy and nutrients but the acquisition and digestion of it will involve energetic costs, the time spent on it cannot be spent upon some other item, there could be costs consequent upon resistance to capture by animal prey, there might be costs associated with the physical or chemical characteristic of the food, and there could be costs due to risk of predation or other hazard during the acquisition and digestion of the food. All of these factors, which are detailed below, must be considered at this time, just as they must during food finding behaviour.

Assessing Energetic and Nutrient Value

A redshank which has arrived at a point on a mud flat where it can see and reach two holes inhabitated by Nereis is likely to choose the larger worm. Such a worm provides more energy, once digested, and is not much more energetically costly to catch. Similarly, a monkey will choose the larger of two fruits and a dog will pick up the larger of two pieces of meat. If the redshank sees a *Nereis* burrow or the burrow of the amphipod crustacean *Corophium* it choses the *Corophium* despite the fact that the net energy return from catching and eating *Corophium* is considerably less than that from *Nereis*. It must be presumed that a *Corophium* diet is more efficient than a Nereis diet as a means of obtaining an adequate balance of energy and nutrients but there is very little direct evidence concerning the mechanisms of such dietary balancing. A similar example is Smith's finding that howler monkeys (*Alouatta palliata*) can assimilate much more energy from fruit than from leaves but they eat both. Again we assume that they can thereby obtain a balanced diet.

The digestive functioning of animal species has evolved at the same time as have the mechanisms for food preference and the ability to learn about dietary requirements. Monophagous animals can obtain all necessary nutrients from their single food source but polyphagous animals, because of their digestive specialisations, usually need a variety of foods, for example, man and the rat must obtain, respectively, nine and ten amino acids from their food in order to survive. There are mechanisms for ensuring that adequate water and energy-giving food are ingested but the only substances which mammals can detect directly in food and ingest when they are deficient in the body, are water and sodium. There is some evidence for a specific calcium appetite in birds. All other balancing of nutrients appears to occur as a result of feeding preferences which are modified by experience.

Animals can learn about diets during development but must also be able to modify food-selection behaviour according to later experience. Young yellow baboons (*Papio cynocephalus*) will sample food which other baboons are eating and will approach individuals which have food in their mouths and sniff the food. The food choices of the baboons are thus influenced by the choices of the other members of their group and this provides a method of avoiding the necessity to sample harmful foods as well as ensuring that the individual is aware of the existence of enough foods to make a balanced diet possible. It does not, however, result inevitably in the consumption of a diet which is optimal in its composition. There must be much variation amongst individuals in diet, that is to say too individuals may not come to the same decision when confronted with a potential food item.

If an individual is suffering from a dietary deficiency it can modify its behaviour in two ways in order to remedy this situation. Firstly, it can refuse to eat the diet which has resulted in the malaise which it presumably feels. *Rozin* found that rats fed on a thiamine deficient diet refused the diet after a few days, i.e., they treated it like a position and took a new diet preferentially. Secondly, the individual can learn to eat diets which remedy deficiency symptoms. *Garcia, Ervin, Yorke* and *Koelling* found that preference for diets containing thiamine was enhanced when the diet was consumed by thiamine-deficient rats. The result of such mechanisms is, as Richter demonstrated with rats, that individuals which have a variety of foods available to them, have the ability to compensate for the variation in nutrient needs which results from pregnancy, lactation, thyroidectomy etc.

Assessing Costs of Acquisition and Digestion

The energy required to acquire food once it has been found may be considerable and in some species it is very much greater than the energetic costs of searching. A cheetah which sees an antelope, or a baboon which sees a coconut or ostrich's egg, still has much to do before it gets a meal. The oystercatcher (*Haematopus ostralegus*) obtains the flesh of the bivalve mollusc *Scrobicularia plana* by probing down the hole in the mud made by the molluscs siphons, grasping the shell, walking in a c'rcle whilst pulling and then opening the shell valves.

In studies of wolves preying upon ungulates, reviewed by *Mech*, the majority of the animals killed were found to be young, or old, or to have some disability. The most likely explanation for this is that these are the easiest animals to catch. In Royama's studies of feeding by great tits the birds were found to concentrate on slow-moving species of spiders, flies and sawflies. Prolonged chases of active prey are energetically expensive and often may be fruitless. This is something which many young animals learn during their development. The experience gained from previous attempts to acquire food items is a major factor affecting many decisions about whether or not to try to acquire a particular item. There have been very many studies of the improvement of various skills when food is a reinforcer, for example, rats or pigeons on complex schedules in operant conditioning experiments, finches pulling strings and chimpanzees using tools. The selection of leaf rather than stem or dead matter by grazing animal is probably related to costs of chewing and digestion as well as to nutrient content. It is likely that digestion costs are a factor which influences decisions about whether to acquire any particular food item but there is little clear evidence for this. The evolution of the mechanisms which result in food preferences must have been affected by digestion costs so that those items which are preferred are usually readily digestible by the individual. Indeed the digestive mechanisms themselves have evolved so that some animals, such as baboons, require low bulk-high-energy food, whereas others thrive on a high-bulk-how-energy food, e.g.., colobus monkeys.

In a choice situation where two items of equal energetic value and with equal acquisition costs are available, optimal foraging hypotheses predict that the item whose digestive costs are least should be chosen. Even if the energy costs of digestion or acquisition are similar for two items, if one item takes longer to acquire or digest it may be avoided. The importance of time as a resource was emphasised by *Schoener* who pointed out that some animals might be time minimisers rather than energy maximisers in their feeding strategies. An animals might have to wait for a long time before a bivalve mollusc opened widely enough to allow its capture. In this case the animal might decide to leave that molluse and rather than wait, attempt to find other food. If a type of food takes a long time to digest, energy return in a given time might be reduced because of the long period when the gut is full.

Exogenous Factors

Food Preferences

All animals are selective in choosing food in their natural habitat, some more so than others. No two species living together at the same time and place at exactly the same staple food. This is the competitive exclusion principle. So feeding selectivity is a subject of major biological importance, not only because it has a bearing upon studies of nutrition, but also because of its relationship to problems of interspecific competition. When animals are given a choice they show preferences. "Cattle exhibit preferences not only for certain plant species but for the same species at different stages of growth, and even for the various parts of an individual plant and for individual plants within a species." The preferences of seed-eating birds, rats and primates are equally clear.

The mechanisms underlying such preferences are not well understood. In seed-eating birds ability to husk a seed quickly and efficiently may determine what seeds are selected. The choice of particular seeds for food correlates well with the size and structure of the bill as in, for example, the Galapagos finches. Experience with different kinds of seeds leads each species to concentrate upon the seeds with the largest and most nutritious kernel which it can efficiently handle. Body size and locomotor ability may play just as important a role as structure of the actual feeding equipment. But within a species a strong element of individuality has been noted by many workers, indicating that a wide latitude in selection of food items is possible.

External Stimuli and Feeding Behaviour

What is the role of external stimuli in the selection of food and in control of the structure of the feeding bout that ensues? Some external stimuli tend to stop feeding behaviour, encouraging rejection or even vomiting. Features such as texture, hardness, and shape may playa part, but chemical characteristics, both smell and taste, are especially important.

Cattle reject food with a bitter flavour, and grasses with a high coumarin content are often avoided. Food soiled with feces deters feeding in many species. Among phytophagous insects the selection of host plants is often controlled by the presence or absence of inhibitory and often lethal chemicals.

Visual characters are sometimes equally important in discouraging feeding. The 'warning colouration' of many animals, various conspicuous colours and patterns which repeal predators, are a good example. These features are often coupled with a noxious taste, smell, or texture, or an effective weapon such as a sting, and predators learn to avoid the warning colours by experience. Sometimes harmless species mimic noxious ones and thus avoid being preyed upon. Certain colour patterns, such as the eye spots on the wings of moths, will even deter a naive predator.

The effectiveness of inhibiting stimuli on feeding depends on the state of the animal. It is minimal after long periods of food deprivation and maximal after a bout of feeding. Competition between eliciting stimuli of varying effectiveness may also effect the outcome. Thus, although repulsive stimuli are significant in determining food preferences (e.g., in rats, rejection of certain food objects in favour of other may also occur through a variety of other mechanisms, with availability playing an important role.

Facilitating stimuli are not normally thought of as having inhibiting properties. Yet there is a sense in which such effects can accrue when the stimuli are encountered in a certain temporal pattern either by loss of the power to elicit responses or by more positive inhibitory effects.

External stimuli facilitate feeding behaviour in several ways. An animal may be sensitized to food stimuli in such a way that contact with one stimulus lowers the threshold of responsiveness to others. Olfactory stimuli often have sensitizing effects, and other senses may work this way as well. In gregarious species social stimuli may facilitate feeding. The presence of another chicken causes a hen to eat more grain, and properties of the food, such as the size of the pile, have a similar effect. Other typically gregarious species such as certain schooling fish, rats, sheep, and primates also change their feeding habits and eat more when they are with other members of the same species.

The eliciting function of stimuli is that most commonly investigated by behaviourists and physiologists. External stimuli also orient responses in space, an important fact often overlooked in laboratory studies of feeding behaviour. The range of effective eliciting stimuli is broad in some species and narrow in others. But we still have astonishingly little comparative information on the specificity of stimuli which elicit feeding behaviour. Both quantitative and qualitative properties of eliciting stimuli are important. The concentration of compounds that can be tasted in food, for example, is an important variable. Several experiments have shown that the effectiveness of substances in eliciting drinking rises with increasing concentration in a solution and then declines. These parallels do not imply that the same physiological processes are involved. The rejection of strong solutions of saccharin is probably a response to its bitter taste. The more frequent rejection of strong salt solutions, on the other hand, seems to be at least partly related to the dehydration resulting from its ingestion.

An example of the qualitative properties of external stimuli that are important in sensitizing and eliciting feeding behaviour is found in the larva of the diamond-backed moth, *Plutella maculipennis*, which responds only to the mustard oil glucosides found in certain cruciferous plants. Chemicals play a prominent role as facilitating cues in plant-eating animals but vision can be significant. Aphids, for examples are attracted to leaves from a distance, distinguishing them from blue sky by a preference for light with a wavelength above about 500 mμ. Vision is especially important to predators.

The varying valence of qualitatively different food stimuli in evoking feeding provides a potential tool for exploring the motivation of feeding behaviour. If we consider eliciting stimuli as either strong or weak, their effectiveness in eliciting feeding should vary with changing motivation to food. In the interval following a bout a feeding only the strongest eliciting stimuli will induce feeding. As the interval since the last meal increases, strong stimuli are more likely to elicit feeding and eventually weak stimuli will become effective as well. In the extreme case, after extended food deprivation, even stimuli which would never normally be effective may acquire eliciting properties. In these circumstances the effectiveness of inhibitory stimuli will decrease. Once feeding has taken place the cycle will begin again. Such an experiment might provide an ideal example of rhythmical changes of responsiveness accompanying a cyclic pattern of behaviour.

FEEDING BEHAVIOUR OF FLIES

Files have much the same basic pattern of feeding behaviour as mammals. Given free access to food, the pattern is typically cyclic, with periods of feeding activity alternating with longer intervals without feeding. The combined efforts of a number of investigator have provided a comprehensive picture of the physiological basis of this behaviour which has great intrinsic interest. In addition to the parallels with mammalian feeding behaviour there are certain striking contrasts.

External Stimuli

The role of external stimuli in triggering feeding behaviour in flies is well worked out. Feeding responses are elicited by a limited class of compounds, primarily certain sugars but also certain proteins. Contact chemoreceptors on the tarsi are stimulated. This sensory input results in extension of the proboscis. Extension brings the chemosensory hairs on the aboral surface of the labellum into contact with the sugar.

In response to this stimulation the labellar lobes open, thus bringing the receptors on the oral surface into contact with the sugar. Stimulation of these receptors, as well as of the labellar hairs, initiates sucking. Feeding is thus initiated and driven by input from oral receptors.

Receptors within the pharynx or esophagus may also be involved. There is some evidence of various groups of chemoreceptors with slightly different functions, perhaps having somewhat different thresholds from one another. In addition to eliciting feeding behaviour, chemical stimuli may also orient the fly to food. Alcohol attracts flies from a distance. Whether chemical stimuli can sensitize flies to eliciting stimuli is uncertain and perhaps unlikely, in view of *Evans* and *Barton Browne's* demonstration that removal of the antennae and palps, which bear the olfactory receptors, does not influence the threshold of responsiveness to beef liver.

Continuation of the Feeding Bout

Once feeding has been elicited it continues for a certain time and certain rate. Both the rat and the duration of the bout of feeding are controlled by external stimuli. Rate seems to be a function of the intensity of the eliciting stimulus, increasing with the sugar solution concentration but declining with the strongest solutions of all. In addition to taste, viscosity limits the rate, especially at the highest concentrations of sugar solutions.

The duration of a given feeding bout is a function of the time taken for the contact chemoreceptors to become adapted to the eliciting stimulus. Consummation is thus the simple result of loss of responsiveness to the eliciting stimulus resulting from continued exposure to it. The stronger the stimulus, the longer the bout of feeding. The initial threshold of responsiveness of the animal is an additional variable to consider. Evans and Barton Browne measured feeding bouts lasting on the average from 51 to 133 seconds, depending on the length of time the animal had been deprived of food.

Feeding Thresholds

After a bout of feeding, the threshold of responsiveness to eliciting stimuli remains high for a time and then gradually declines to a level at which feeding will occur again. The share of the recovery curve depends on the kind of sugar used. The threshold declines somewhat 20 minutes after the end of the previous bout, probably because the effects of adaptation are waning. Then the threshold rises again to maximum before declining once more. This second recovery phase lasts too long to be related to sensory adaptation and suggests a mechanism differing from that involved in the initial threshold decline. This conclusion is confirmed by the discovery that sugars which normally do not elicit feeding and are thus assumed not to be tasted by the tarsal receptors have the same effect of changing the threshold of responsiveness to glucose for a long time. Some consequence of feeding has resulted in a prolonged change in the threshold of responsiveness to eliciting stimuli.

Feeding thresholds are unrelated to the nutritive value of assimilated food. Ingestion of sugars which have no nutritive value for

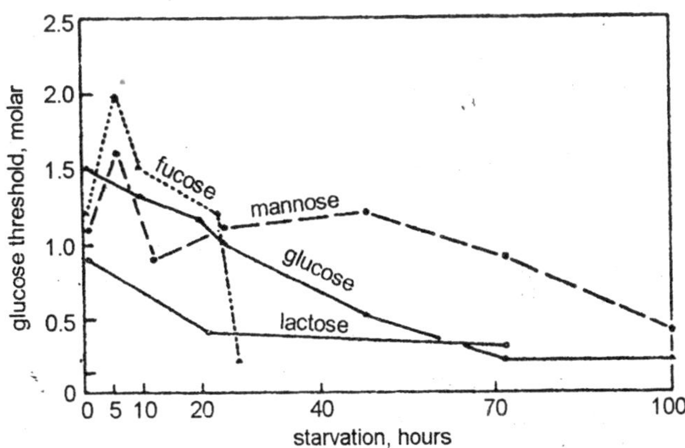

Fig. 2.6. The minimal glucose concentrations for eliciting feeding in flies after varying periods of starvation, following feeding to saturation with four different sugars.

flies, such as fucose, for example, still causes elevation of the feeding threshold. Injection of nutritive sugars into the haemocoel of a starved fly has no effect on the threshold. Furthermore, threshold changes are not related to the timing of changes in the level of normal blood sugar, trehalose. Nor is the simple performance of the motor activity of feeding responsible for satiation, since the detached head of a satiated fly will feed. Experiments involving delicate surgery have localized the origin of inhibition of feeding somewhere in the foregut other than the crop. How the effect is mediated is still uncertain. Evans and *Barton Browne* suggest a possible hormonal link. Chemoreceptors in the wall of the foregut may also play a role.

Locomotory appetitive behaviour for he feeding is closely related to changes in the threshold of the feeding response.

Foraging Behaviour

Food finding by hummingbirds is facilitated by the advertisement of the food source by the plants. The flowers are rendered conspicuous to the hummingbirds by their large size and long wavelength colouration. Goldsmith has demonstrated that hummingbirds can see at wavelengths which we call ultraviolet because objects reflecting these wavelengths cannot be seen by man. Flowers usually occur in patches because they are on a large plant or are part of a collection of small plants flowering synchronously. Nectar production often continues for several days. Patch finding might be energetically costly in some species, such as those which live in the mountainous areas, but hummingbirds must find the majority of patches within a small home range or territory. Territorial defence in hummingbirds and other nectar feeders is discussed by Carpenter (1978) and Wolf (1978).

The importance of learning the location of patches and the rate of renewal of the nectar supply is greater than that of actually finding the food at most times in the lives of hummingbirds. Foraging behaviour in these birds involves a series of decisions about when to leave a patch and when to revisit a patch. Time spent in a patch of inferior quality could often have been spent in a good quality patch. Before considering such decisions it is necessary to review the information about food intake.

Hummingbirds lick nectar from plant nectaries with their grooved tongues and pass it to the crop. It seems likely that the initiation of feeding depends upon the emptiness of the crop but many other factors must be involved. The size of crop is proportional to the weight of bird. The X-ray studies by these authors show that the food is passed to the rest of the digestive system from the crop, a full crop emptying in 30-40 min in the case of *Eugenes fulgens*. Hummingbirds in the laboratory or in the field usually fly to the food source, be it a feeder or a collection of flowers, ingest enough nectar to half-fill the crop and then return to a perch. This behaviour is repeated with the result that energy is gained at a relatively constant rate during the day.

Feeding Strategies

Possible feeding strategies which could be adopted by a hummingbird are discussed by de *Benedictis* et al. (1978). The birds could attempt to minimise feeding time but this would involve filling the crop whenever possible, not half filling it as is observed. The same prediction follows from the strategy of maximising net energy gain per bout. The model of feeding strategy which fitted the observed feeding data best was that in which the rate of net energy gain during the total period from the beginning of one feeding bout to the beginning of the next was maximised. This model took into account the facts that the energetic cost of hovering increases as nectar is ingested, thus increasing body weight, the return flight requires more energy than the outward flight, and the inter-bout period on the perch also involves energetic cost. *Hainsworth* (1978) has extended this model to incorporate overnight storage effects on energetics. The information required for this model includes body mass, duration of light and dark periods, body temperature, ambient night temperature, rate of energy expenditure in relation to body mass and temperature differential, and the energy lost when converting sugars to fat and back again. An estimate of the amount of energy required for storage can this be obtained. This information can be combined with calculation of energy expenditure throughout the feeding period and energy obtained from the food.

Hainsworth's model predicted that maintenance costs during the day would have a major effect on feeding frequency and this was shown to be correct in laboratory studies by reducing temperature, which increased feeding frequency. The model also predicted that increased costs overnight would be remedied by increasing meal size on the following day and this was shown to occur. The function of these mechanisms must be deduced from consideration of the general biology of the hummingbird. The regulatory mechanism in the daytime is that which is the most efficient for maximising the rate of energy gain from the available resources. If the time available for feeding is restricted, such as during preparation for a long night or a migratory

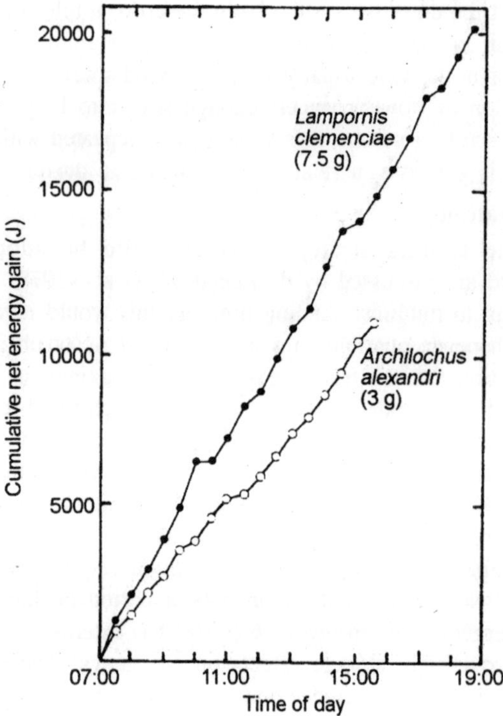

Fig. 2.7. Humming bird species of two different sizes showed approximately constant rates of net energy gain during a day.

journey, the feeding pattern is modified. The meals are larger as part of a feed-forward mechanism which anticipates the greater bodily demands. In situations where the risk is high this food supplies adequate for requirements will not be found, the bird may fill its crop completely.

Another decision which must be taken by a foraging hummingbird is when to give up feeding in a patch. The marginal value theorem of Krebs et al. (1974) and Charnov (1976) states that the animal will leave a patch when its net rate of energy intake has dropped to the average net rate of energy intake in the entire habitat. Pyke (1978b) suggests that in order to do this a hummingbird should leave a patch when its estimate of the nectar volume obtainable from the next flower, divided by the average time required to move to the next flower and remove the nectar, is less than its overall rate of nectar intake in the patch. Observations suggest that hummingbirds use information about the number of flowers in the patch, the number probed and the amount of nectar in the present and the last, or last two, flowers in deciding

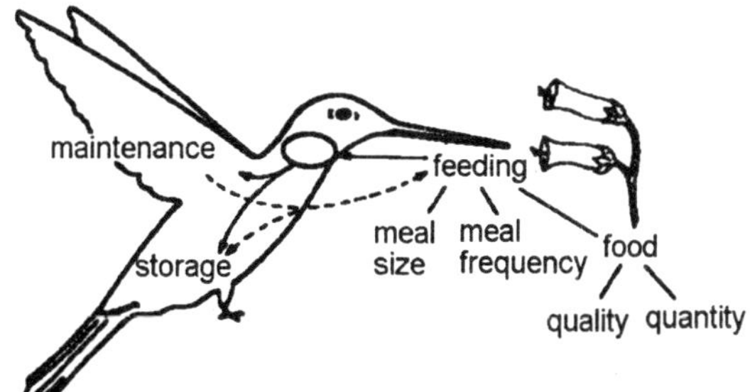

Fig. 2.8. Humming birds consume nectar which is collected from flowers at intervals and stored briefly in the crop before being used for body maintenance or converted to a food reserve, usually fat, for overnight maintenance.

whether to depart. Although Pyke's model explains his data, it would seem likely that where patches are widely scattered, an estimate of the energy required to move to another patch might be a factor affecting the threshold level of intake at which departure occurs.

Most of the studies of hummingbirds foraging which have been mentioned are of situations where nectar is abundant and feeding behaviour is not much affected by predation risk or by competition for food, nest-sites or mates. Studies of hummingbirds living in mountainous areas where nectar supplies are often sparse and the hummingbird assumes a torpid state during the cold nights, show that vigorous defence of feeding territories occur and community structure is complex. In such competitive situations, foraging behaviour can be inextricably interwoven with other functional systems so that discussions of optimal foraging alone do not go far toward explaining the behaviour which is observed.

Feeding in Cattle

The majority of studies of animal feeding in relation to energy utilisation have been carried out on farm animals and some ideas about feeding strategies have arisen as a result of this work. Other ideas have resulted from work on laboratory animals by physiologists or phychologists and work on wild animals by behavioural ecologists (e.g., the study of moose by *Belovsky* (1978) and Belovsky and Jordan (1978). The interchange of such ideas is desirable since many of the same fundamental mechanisms operate the farm, in the laboratory or in the wild. Cattle are ruminants which regurgitate their food from their multicompartmental stomach and chew it during the resting phase after eating is completed. Much of our knowledge of ruminant nutrition comes from indoor feeding studies but field studies are of great importance for world food supplies. Less than 5% of the world's domestic ruminants spent a significant period indoors in 1970 (F.A.O. 1970). The foraging behaviour of cattle and sheep, for example the detailed studies of the late *T.H. Stobs* (e.g. *Chacon* and *Stobbs* 1976) and *G.W. Arnold*, can be related to information about energetic efficiency. Such studies, therefore, aid in the management of farm animals as well as providing an example of feeding cost-benefit analysis for behavioural ecologists and physiologists.

For many animals, the problems of food-finding limit energy input for ruminants, food-processing takes a long time and is often the limiting factor. The gut usually has food in it so digestion, aided by symbiotic bacteria and protozoans, is a continuous process. The upper limit of the overall rate of processing is determined by the cross-sectional area of the gut and the meal must end once the rumen is full. A consequence of the need for large quantities of food and of slow food processing is increased duration of feeding time and of digestion time. A high proportion of the life of animal, therefore, is devoted to these activities. Wild ungulates, however, also have to find patches of grass or other vegetation upon which to graze or browse. Most wild and domestic ungulates compete with conspecifics for the food which is available. The best way of competing is, usually, to eat as fast as possible when food is found and this behavioural strategy is at least as important to cattle whose grass allowance is carefully calculated, as it is to wild ungulates. Another important objective of the feeding strategy in large generalist herbivores is to obtain the best mix of nutrients within a fixed total intake.

Grazing Decisions

The existence of such behavioural components in the feeding strategy is obvious to the behavioural ecologist but is sometimes overlooked by those concerned with animal husbandry. Cows do not eat randomly all green material which they encounter, as is emphasised in the next section. They have to make a series of decision, when grazing which will differ according to the characteristics of the pasture upon which they are feeding. These decisions include whether to lower the head and bite the herbage, how large a bite to take, at what rate to bite, whether or not to stop biting and chew or otherwise manipulate the grass in the mouth, whether to swing the head to the side and how far, whether to take one or more steps forward and whether to raise the head and carry out some other behaviour. The long-term consequences of such decisions are that total daily grazing duration is

4-14h at bite rates of 40-8- min^{-1} and with bite sizes of 0.05-0.7g. Each day, the food intake may be up to 12.5 kgs dry matter and the cow will ruminate for 4-9h (*Hafez* and *Dyer* 1969, Stobs 1974b, Chacon and Stobbs 1976). In temperate countries cattle feed most in the daytime. Visual food selection is then possible but diurnal. Feeding may be a relic of anti-predator behaviour for animals can feed and keep looking for predator in the daytime but it is safer to remain immobile at night.

Selection of Grazing

The fact that cattle show feeding preferences is apparent when a grazed field is examined. The sward is less uniform than it was before grazing. Some plant strains and species, usually the more digestible, are preferred to others and leaves are eaten in preference to stems. Some preference studies indicate avoidance of high concentrations of secondary plant chemicals, e.g., tannin, coumarin, alkaloids etc. *Donelly* (1954) found preferences by cattle, among 1206 strains of *Sericea lespedeza*, for low tannin content and fine stems. A variety of other studies has shown clear relationships between alkaloid levels and food preferences of sheep (review by Arnold and *Dudzinski* 1978). Some ruminants, however, have digestive adaptations which allow them to eat plants with alkaloids or other poisons.

Many other studies have related nutritional properties of plants to the extent to which they are eaten by ruminants. There is, as yet no evidence for long-delay learning of nutrient quality by ruminants but the list of deitary requirements of cattle shown in Table 2.1 (A.R.C. 1965) emphasises the importance of cattle's feeding preferences and the likelihood of occurrence of subtle modifications of diet with experience. It is known that prolonged grazing of grass low in cobalt leads to extreme reduction of intake, which would presumably be associated with movement to a new feeding area it this were possible.

Man limits the choice of food plant for many cattle but plants different stages of development and those contaminated with dung, which is avoided, are still encountered. The composition of pastures is altered by the feeding activities of cattle, by their dung and the fact that they tread heavily on plants as they move around the pasture. Plants which die when grazed or trodden become rarer, whereas those which are avoided become commoner. When the shoot of a grass plant is eaten to within 5 cm of the ground the next shoot is more likely to grow along the ground, so grazing results in a smaller proportion of upright shoots and a larger proportion of horizontal runners and procumbent leaves. Plants near dung pats can survive with a more upright growth form than those in other areas.

Evolutionary changes in anatomy, physiology and behaviour of grazing animals have been a response to the characteristics of grasses but, in response to the grazers, changes in grasses have also occurred. The occurrence of secondary plant chemicals has been mentioned above but the growth from and the mechanical properties of grass must also have changed. The arrangement of fibres in grass in such that cows have to work hard in order to break off the leaves.

Table 2.1. Daily dietary requirements of cattle

Nutrient	*Requirement for 400 kg heifer growing at 0.5 kg day^{-1}*
Calcium	25.8g
phosphorus	23.7g
Magnesium	6.8g
Sodium	7.5g
Potassium	(usually adequate in diets)
Chloride	10.9g
Iron	0.24g
Copper	0.8g
Cobalt	0.0008g
Zinc	0.42g
Manganese	0.32g
Iodine	0.001g
Molybdenum	(Trace)
B carotene	0.032
Vitamin D	100 international units
Protein	210g if low fibre diet
Protein	360g if high fibre diet

Input and Output of Energy

Intake by grazing animals can be determined by observing behaviour, recording movements automatically, using an oesophageal fistula, estimating dilution of an indigestible substance, or measuring grass before and after grazing.

The potential energy obtained from plant material can be determined by calorimetry. If a known amount of food is consumed by a cow, the metabolisable energy of the food can be calculated by subtracting from the gross energy obtained form it, the energy obtainable from the faeces, urine and fermentation gases. The metabolisable energy provided by pasture is 7-14 kJ g^1 dry organic matter. The metabolisable energy may be used for body maintenance, growth or milk production and some is lost as heat. On a high quality diet, 76% of metabolisable energy can be used for body maintenance, assuming the there is no growth or lactation. Additional food above that for maintenance can be used with efficiencies of 62% for growth and 64% for lactation. Interindividual comparisons of energy utilisation must take account of body weight. The logarithm of metabolic rate when fasting divided by the logarithm of the body weight is usually close to 0.75 so the scale factor for body weight which is used in comparative energy studies is weight 0.75.

Energy expenditure can be estimated by measuring respiratory gas exchange. In the laboratory this can be carried out by enclosing the animal in a respiratory chamber and measuring oxygen consumption and carbon-dioxide production. Alternatively, inspired air flow can be monitored and expired gases can be collected by means of a tube connected to the trachea of the animal. Neither method is very satisfactory for the metabolic processes may be altered by the treatment, but both of them allow some estimation of the energetic costs of different activity. Indirect estimates of daily energy consumption, by continuous infusion of radioactively labelled sodium bicarbonate and sampling of blood or urine, are similar to those obtained by gas exchange methods. Young (1970) found that young (12-14 months) grazing heifers used 706 KJ $kg^{-.75}$ $24th^{-1}$. The results from such studies give values for maintenance energy 20-70% higher for grazers than for animals housed indoors. The total daily energy expenditure estimates are accurate enough to allow the evaluation of particular foods and predictions about requirements. The assessment of the efficiency of feeding strategies, although well worthwhile, is more difficult because results such as those shown in Table 2.2 are variable. External conditions affect costs, e.g., a 10 km h^{-1} wind increased oxygen consumption of sheep by 40%, but sophisticated energy monitoring methods will soon make it possible to access the energetic costs of the various components of feeding strategies.

The Control of Grass Intake

The physiological mechanisms controlling food intake by ruminants are reviewed by Arnold (1964) and by *Baile* and *Forbes* (1974). The initiation of grazing by cattle often occurs at particular times during

Table 2.2. Energetic costs of different activities by cattle and sheep

Activity	*Cost*
Cattle	
Lying (after 10 mm +)	422 KJ $kg^{-0.75}$ $24th^{-1}$
Standing	290 KJ $kg^{-0.75}$ $24th^{-1}$
Lying—standing	0.66 KJ kg^{-075}
Standing—lying	0.36 KJ $kg^{-0.75}$
Walking	2.01 $kg^{-1}m^{-1}$
Sheep	
Grazing	2.25 J $kg^{-1}h^{-1}$
Ruminating	1.00 J kg^{-1}h-[1]

the 24th period and behaviour of individuals is affected by that of others in the social group. It is likely that information from visual, olfactory and gut receptors and from fat depot monitors are factors which affect the start of grazing. The termination of a meal is not normally limited by oropharyngeal factors, for animals whose ingested food is removed via oesophageal fistulas consume much more than they do during normal feeding. When Campling and Balch filled the rumen with ingesta from another animal, however, the meal was terminated early. Water or air-filled balloons had the same effect. Removal of rumen contents extended the meal length unless the intake rate was low because the pasture had already been well grazed. As a consequence of such experiments, most authors agree that when the volume of rumen contents reaches a certain level, receptors detect the distension and the meal is terminated.

Since blood glucose levels decline during feeding and injection of glucose into the body does not reduce the rate of feeding unless the dose levels are very high, it is unlikely that glucose levels are of much importance in terminating feeding. Stimulation of the lateral or ventromedial hypothalamus has similar effects in ungulates to those obtained in rats, so these areas are probably involved in the control of feeding, but no gold thioglucose lesions can be produced in ungulates. Baile and Forbes (1974) consider that input from fat-depot monitors may be important in the control of feeding but the evidence is not clear. They also propose that input from monitors of intestinal content would be more efficient in long-term control of eating than would rumen distension monitors since the latter would provide a poorer estimate of true intake.

Behaviour of Grazing

The grazing strategies of cattle must include modifications in grazing and ruminating behaviour which depend upon gut contents, upon the quantity and quality and of herbage available now and upon that predicted for future. Domestic cattle have to contend with open range situations, where patches of adequate grazing are widely separated set stocking, where a herd of cattle is kept in a field whose grass renewal rate is sufficient to supply their needs and rotational grazing, where they graze on a paddock or strip of pasture for 1-10 days and are then moved to another. The amount of food available may be barley sufficient for body maintenance or may be enough to allow the production of large quantities of milk. The grass diet may be supplemented with high quality concentrates. These extremes may be encountered within a period of a few days by animals on a rotational grazing system.

When a group of six cows were put into a Setaria pasture in Australia for 14 days by Chacon and Stobbs (1976) they showed initially a very clear preference for leaves but as the total amount of herbage available declined they ate more stem and dead material. The duration of grazing was quite high (over 9th) at the beginning of the 14 days but increased to almost 11th on days 3-6, and then declined. In other studies, grazing time has been shown to increase considerable from lower initial levels as the amount of available herbage declines. In Chacon and Stobbs' experiments the rate of biting increased during the first eight days but the bite size declined throughout the 14 days. The estimated intake dropped a little between the first two and the second two days, despite behavioural compensation and dropped considerably after that. These results show that grazing behaviour is modified in a way which is energetically costly but which increases intake when the quality of the food is good. The most costly grazing behaviour is not used, however, when the quality has declined. The previous experience of cows in rotations grazing experiments must alter the likelihood that they will expend a lot of energy when the grass has been grazed down. Many cows are quite capable of training farmers to move them to a new paddock when available herbage is low. The cue which they train the farmers to use is the sight and sound of a row of cows standing by the fence and bellowing.

In addition to the rate of biting to the rate of biting and the bite size. Cows can vary manipulatory movements, the amount of walking and the amount lateral head movement which they carry out during a feeding bout. A comparison of the same animals grazing on long grass, i.e., mean length of longest part of shoot 30cm, and short grass, i.e., mean length 13 cm, showed that total grazing time was 7.9h per day on short grass but 6.9h on long grass. The mean times spent walking were 56 min and 30 min respectively. When the cows were videotaped whilst grazing, the rate of biting decreased by 8% and the distance walked whilst grazing decreased by 24%. There was an increase of 27% in the number of manipulatory chews. The bite size of cows which had fasted for as little as 16th was about 10% larger than that of cows which had fasted for only 2h but most variation in bite size and rate of biting is related, principally, to the fact that cows will search for leaves and will take them individually in poor pasture.

Another factor which affects grazing behaviour is the presence of faeces on the pasture. The dung from one cow, without decomposition, may cover as m..ch as $200m^2$ in a grazing season and this may affect 10-15% of the pasture required in a season. Cattle will avoid the area close to dung-pats and will preferentially eat clean herbage rather than herbage treated seven weeks earlier with slurry from a cowshed. On slurry-treated pasture cows ate less unless they did not have to eat the grass down to the slurry. If slurry was present they were more likely to stop grazing and walk a few steps and were involved in competitive encounters more frequently than on clean pasture.

Grazing behaviour is also affected by weather and by disturbance. Grazing ceases in driving rain or if a large predator is heard or smelled. The sight of a dog or the occurrence of any unfamiliar activity on the farm may result in an interruption of grazing behaviour, probably as part of an anti-predator response. Calves are more likely to be disturbed by any particular event, perhaps because cows have had much previous experience of such events which had no adverse effect on them. The development of grazing in calves involves an increase in grazing time during the first four months as rumen function develops. By one year of age, calves graze for an hour a day more than heifers and for 1.6h per day more than 3½ year-old cows. This may be because calves are more selective size or experience of grazing has improved the feeding efficiency of the older animals. *Arnold* and *Maller* (1977) and Arnold and Dudzinski (1978) showed that sheep reared for three years with no grazing experience sheep and also reported that preferences for pastures plants were much affected by previous experience.

Anti Predator Behaviour

As most species are preyed upon by animals of at least some species, avoidance of predation plays an important role in survival and reproduction. Some defensive mechanisms operate throughout much of

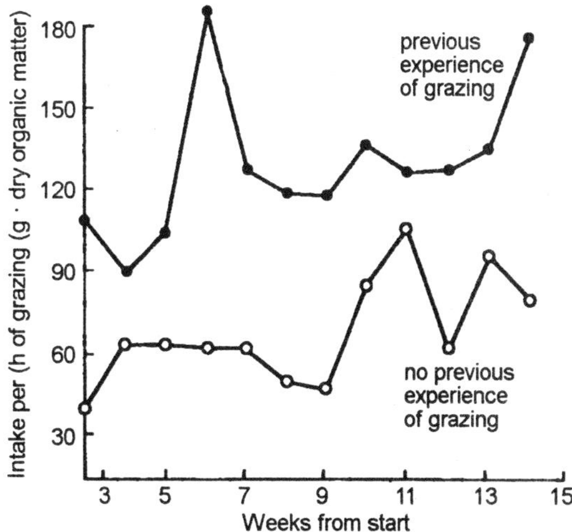

Fig. 2.9. Three-year-old sheep which have had no experience of grazing (open circles) are less efficient grazers than are sheep which have grazed (filled circles).

the life of an individual whereas others are used only when a predator is detected or when a predator attacks the individual. *Edmunds* (1947) refers to two types of mechanism, primary and secondary. Following *Kruuk* (1972) Edmunds define primary defence mechanisms as those which operate regardless of whether or not there is a predator in the vicinity. The behavioural components of such mechanisms may occur or before any predator is detected. The mechanisms include (1) hiding in holes; (2) the use of crypsis; (3) mimicking inedible objects; (4) exhibiting a warning of danger to predators; (5) mimicking individuals in category (4); (6) timing activities so as to minimise the chance of detection by a predator; (7) remaining in a situation where any predator attack is likely to be unsuccessful because of possibilities for secondary defence; (8) maintaining vigilance so as to maximise the chance of detecting the advent of a predator. All of these mechanisms decrease the chance that a predator attack will occur. Many of them have a considerable influence upon habitat selection. For some species which are very vulnerable to predation, the avoidance of predators limits the distribution of the animals even more than do factors related to body maintenance and feeding. The selection of a suitable habitat may be accomplished by the recognition of a factor which is not in itself advantageous but which is an indicator of the presence of advantageous features; such as those which make possible feeding or predator avoidance. For example, a bird, such as nightjar, which is camouflaged when amongst dead wood might fly over open downland towards a dark green mass of woodland because suitable hiding places are likely to be found around the edge of the wood.

Secondary defence mechanisms are defined by Edmunds ad those which operate during an encounter with a predator. The encounter may just involve the supposed or actual detection of a predator by the prey or it may involve an attack by a predator. Some secondary defences are antirely pesive. The prisim-spined sea urchin *Diadema* or the thick-skinned rhinoceros are adequately protected from attack. No behavioural specialisation is needed. Active forms of secondary defence include (1) exaggerating primary defence, e.g., a camouflaged animal remaining motionless; (2) withdrawal to a safe retreat; (3) flight; (4) use of a display which deters attack; (5) feigning death; (6) behaviour which deflects the attack to the least vulnerable part of the body, to an inanimate object, or to another individual; and (7) retaliation.

Both primary and secondary defence mechanisms are used by most animals and different mechanisms are often used for different predators. In an encounter with a predator, the prey animal may be able to use several alternative anti-predator behaviours. Decisions must be taken about, for example, whether to remain motionless, to flee, to display, or fight. If one method is tried, without successful avoidance of attack, others may be tried. There are many possible strategies which involve alternative courses of action depending upon the characteristics and behaviour of the predator.

Defensive Responses by Farm Animals to Man

To the ancestors of present-day farm animals, man must have been recognised as a dangerous predator. Despite many generations in domestication, the anti-predator functional system still has considerable effects on the behaviour of farm animals. In some situations, predation by wild animals or dogs is till possible but anti-predator behaviour now is directed against environmental changes who origin in largely unknown to the animals, or against man. The occurrence of unexpected sounds or the arrival of olfactory or visual portents of danger may elicit adrenal and behavioural responses which are the same as those shown by their wild ancestors. The detection of a person who remains at a considerable distance from the animal may elicit no such responses but the close approach which is often necessary on farms must often elicit a defensive response. The fact that people are often treated as predator by farm animals is ignored by many of those who work on farms on who attempt to understand the behaviour of the animals but the competent stockman has learned observation about such behaviour. Good stockmen also know how to minimise the anti-predator behaviour which is directed at them, for if they can do this they can often improve production and enjoy their work more.

The importance of the abilities of the cowman in obtaining the best possible production from a dairy herd was stressed by *Albright* (1971). Milk let-down occurred faster if the cow saw the familiar milker. Yield was higher if the cowman was delioerate in movements, talked quietly to the cows and followed a regular routine with no panic activity. Other effect of a 'good' cowman were that cows entered the milking parlour, and were more approachable in the field. It seems likely that a 'poor' cowman elicits a greater adrenal response and corresponding defensive behaviour so that management of the dairy herd is more difficult and milk production is impaired. Studies of the defensive responses of cows to cowmen and to disturbing aspects of yard and milking parlour design, are helping to improve management and milk production. The detection of a person in an animal-house may be a traumatic experience for comparatively small animals such as chickens or turkeys. If a person enters a house rapidly and noisly the nearest birds may show a violent escape response. Whether the birds are in separate cages, or in one large aggregation on the floor of the house, a wave of violent escape behaviour, sometimes called 'hysteria', may be transmitted down the house. Such behaviour can be minimised by a quiet knock before entry and by slow, quiet movements while entering and when within the house. The economic importance of quiet slow movements in poultry and other livestock houses is obvious to the turkey farmer who has found the corpses of dead birds under a heap of turnkeys disturbed by the entry of a noisy person. The effects of defensive behaviour and adrenal activity on larger animal may be obvious, for example, if a fleeing cabreaks a leg, but many effect are less obvious. Intermittent high adrenal activity may affect adversely milk or egg production, growth rate an disease resistance as well as ease of handling.

Warning Colouration and Mimicry

Some animals protect themselves against being eaten by containing poisonous or sickening substances. Some such animals make their own poisons; others takes them from sources in their environment. The wings of the monarch butterfly, for example, contain powerful heart-stopping poisons called cardiac glycosides. The monarch eats the poisons as a caterpillar, when its food plant is the asclepiad, or 'milkweed', which contains cardiac glycosides. The caterpillar is not harmed by the poisons; it just stores them, and they are then retained by the adult.

For the behavioural problem of defence by poison, we must turn from the prey to the predator. If the defence is to work, the predator must learn not to eat poisonous animals, because natural selection will not favour a trait, by which an animal, after it is dead, makes its attacker sick. If the trait is to evolve, it must ensure survival. The tactic used is to enable predators to learn to recognize sick-making prey, in which case they will avoid them. It is a skill predators will readily learn. When a bird eats a monarch, as an adult or caterpillar, it will be violently sick within minutes, an experience it would learn not to repeat. The bird's problem then is to distinguish sickening from edible prey. Now, the colours of the monarch are bright gold, and ethologies suspect that the bright colouration evolved to make the monarch more memorable to birds. It is therefore called warning colouration. Experiments have shown that predators learn to avoid sickening prey, J.V.Z. Brower, for instance, offered the monarch butterfly as food to the Florida scrub jay. On their first meal of monarch the jays were violently sick; but after only a few trails they had learned not to eat monarchs, though they continued to eat other, tasty food. What has not yet been conclusively demonstrated is that birds learn to avoid brightly coloured sickening prey more quickly than equivalent but duller coloured prey. It is necessary for the theory that they should, for otherwise poisonous prey might just as well be dull as brightly coloured. Few ethologists would doubt, however, that such an experiment would be successful.

The early stages in the evolution of warming colouration pose a paradox. The population of monarchs clearly benefits from teaching each generation of birds not to eat them. They benefit by being eaten less. The problem is that some butterflies have to be eaten to teach the birds to begin with. In evolutionary terms, it is no consolation to the dead butterflies that some others are benefitting from their death. They are dead; they have failed to reproduce; natural selection has worked against them. When warning colouration first appeared in evolution, it would have been a rare, minority characteristic. Those rare, brightly coloured butterflies would have been conspicuous to predators, and therefore eaten. One would expect natural selection to have eliminated the characteristic. How could natural selection favour an increase in its frequency? The question has not been satisfactorily answered, though there are some possible answers. One possible answer is that the family relatives of the eaten butterfly may benefit from its death. Members of the same family tend to resemble each other; if one were warningly coloured, many others would be as well. If one member were killed, the rest of the family would benefit from the lesson taught to the predator.

Camouflage

The nochid moth's defence is to seek escape in active flight. The opposite defence is to sit dead still and try to be invisible. Such is the method of camouflage in which a species evolves to resemble its background. Camouflage is of course an adaptation of appearance and colouration, but the most exquisite artistry will be wasted if the animal's behaviour is not suited to the camouflage. The world is a patch-work of different colours: the animal is only camouflaged if it settles in the right place. Consider the European grass hopper *Acrida turrita*. It comes in a green form and a yellow form. In a nature the green form lives in green places and the yellow form in yellow and brown places, with rare exceptions. In a simple experiment, the German ethologist *S. Ergene* gave yellow and green grasshoppers a choice between yellow and green backgrounds. The green grasshoppers fittingly tended to go and settle on the green backgrounds, and the yellow grasshoppers on thc yellow.

The North American moth *Melanolophia canadaria* faces a more difficult problem in tining up with its background. It has striped wings and lives on the bark of trees. It must line its stipes up with the lines of the bark if it is to be camouflaged. In an experiment, *T.D. Sargent* allowed striped wings and lives on the bark of trees. It must line its stripes up with the lines of the bark if it is to be camouflaged. In an experiment, T.D. Sargent allowed the moths to sit on cylinders that had regions of vertical stipes and regions of horizontal stripes. If the stripes (which were made of black tape, struck on a white surface) could be felt by the moths, then the mouths usually lined up correctly. When Sargent covered up the stripes and surface with a transparent film, the moths no longer lined up correctly. The moths must be relying on the feel of the surface that they have to line up on. In nature they will be able to feel the stripes of their background, and ensure that they settle in a camouflaged posture.

Escape

Rapid and agile locomotion provides the best and probably the most common means of escaping from predators. Various species supplement their locomotor escape patterns with displays that function to distract or startle a potential predator. Still others may adopt a state of tonic immobility ("play possum") as a means of reducing the likelihood of attack.

Social Antipredator Behaviour

Confusion Effect

One of the main benefits of sociality is though to be the protection gained, as stated in the adage, 'safety in number'. Living in flocks herds, or schools increases the chance of spotting predators. When approached or attacked by predators, groups usually become more compact. Fish schools, however, usually form a vacuole around a predator. Tightly packed prey make a difficult target for the predator, which must select an individual toward which it can aim its attack. The predator tries to isolate an individual from the group; any individual that strays from the group or is in any way different from the rest is likely to be attacked. For example, when experimenters presented hawks with set of ten mice, the hawks preferred the oddly coloured mouse *Mueller* (1971).

A primary means of protection is the confusion effect, described by Miller (1922). A variety of predators that attack swarming prey exhibit a lower prey capture rate, a longer period of hesitation, and more irrelevant behaviour than when they attack solitary prey. Factors that enhance the confusion effect the swarm number, swarm density, and the uniformity in appearance of swarm members. The mechanism producing the confusion effect is not understood, but a prenator seems to be more easily frightened when it feeds in a dense swarm of prey; perhaps the predators finds it harder to detect other animals that might prey on it. In laboratory experiments with three-spined sticklebacks (Gasterosteus aculeatus) feeding on water fleas (*Daphnia*), fish with experience in feeding on dense swarms fed more efficiently when tested with dense swarms than did those experienced only in feeding on low-density populations.

Detection

We might expect to find groups of mixed species if the combination of different sensory capabilities is advantageous. Baboons and ungulates are frequently together on the African plains. The aboreal habits and keen vision of baboons and the well-developed olfactory systems of ungulates probably increase the distance at which prey can be detected. It is clear that these different species recognize and respond to each other's alarm calls.

Social living increases the likelihood that predators will be detected and allow group members to spend more time in other activities.

Individuals in large, dense colonies prairie drops (*Cynomys* spp.) spend less time in alert postures than do those in small colonies. Once predators are detected, alarm calls alert neighbours to danger. *Marler* (1955) has argued that in some birds the alarm call is difficult to locate because it is a high-pitched note in a narrow-frequency range. The alarm-calling ground squirrels studied by *Sherman* (1977) emitted calls that were easily located, and calling squirrels were more likely to be caught than non-callers. Close relatives of the caller were likely to be close by and thus could benefit from this seemly altruistic behaviour.

Concealment

Many prey species concerned themselves from predators by remaining in protected locations—such as burrows, crevices, or huts—when predators are present. Alternatively, concealment can be provided by the actual appearance of the animal itself. Cryptic colouration, which functions to enable animals to blend in with the background, can be found in virtually all animal taxa. Many extreme examples are found in insects, which may resembles leaves, twigs, or even bird droppings. Often, a particular behavioural pattern is associated with protective colouration such that the animal orients itself in a particular relationship to its environment, usually remaining motionless.

Freezing

Another strategy in response to predators is to freeze. The presence of protective or cryptic colouration is often associated with this behaviour, as in the spotted white-tailed deer fawn (*Odocoileus virginianus*). Some animals carry this strategy further and feign death. For example, the opossum (*Didelphis virginiana*), is harassed by a predator such as a dog, remains motionless on its back. The physiological correlates of this behaviour are not completely understood. Hog-nosed snakes (*Heterodon platyrhinos*) exhibit similar behaviour, although they may precede it by threat behaviour in which the hog-nosed snake resembles a cobra. Freezing and feigning death may work because many predators seem to respond only to moving prey. For example, wolves will not attack prey that stand motionless.

Development of Anti Predator Behaviour

A baboon which is attacked by a leopard but which survives, learns something about the speed and attack methods of leopards. Future encounters with leopards will be different as a consequence of the experience. Field studies of the improvement in techniques of anti-predator behaviour are difficult, but many laboratory studies of avoidance conditioning minic certain aspects of this type of behavioural change. Suppose that a rat in a cage learns to avoid shock by pressing a set of levers in a particular sequence when a certain combination of stimuli is presented. That rat may be using abilities which would also be of use were it faced with successive exposures to a predator. The effect of experience are often easiest to study in developing animals since some aspects of previous experience can then be controlled.

In species in which parental care is prolonged, the most effective anti-predator behaviour which young animals can show is often to recognise and maintain contact with the parents and the place where they are put by the parents. The overriding importance of finding the parent leads to some apparently anomalous responses to predators. *Lorenz* (1935) describes how newly hatched precocial birds, such as goslings, show no avoidance behaviour to a variety of moving objects including, for example, man, a goose predator. Schaller and Emlen (1962) pushed objects towards the young of ten species of precocial birds and found very little avoidance reaction at 10 h of age. Avoidance in their experiments, and in those of Bateson (1964) on chicks exposed to novel moving objects in a runway, increased considerably during the first week of life but changed little after this time. The startle responses shown by young domestic chicks in their home pen, when a light-bulb on the wall was illuminated for the first time, increased in magnitude and duration with increasing age from 1 to 10 days. The change with age of these studies was greatest during the first three days. As the animal gets older, it learns the characteristics of its immediate environment. If the degree of novelty of any environmental change depends upon a comparison with a 'model' of the familiar surroundigns, novelty will increase whilst the model is established and hence responsiveness increases with age.

A system results in the young animal showing anti-predator behaviour to any novel change in its surroundings will operate when a real predator appears. Responsiveness to potential predators by young animals does not depend solely upon degree of novelty. It is also affected by the nature of the sensory input which results from the presence of the predator. Bodily damage, a looming object, or a loud noise will elicit more extreme anti-predator behaviour than the sight of small objects, the sound of quieter noises, etc. The existence of perceptual analysers in the brain which enable an individual to recognise dangerous events, for example, a looming object. Some of these analysers are known to develop in a different way according to specific visual experience by others probably depend on general environmental factors such as light and metabolic stall at key times during embryological development. The average individual, therefore, would have some predator detection mechanism whether or not there had been previous visual experience of the predator. An elaborate and specific possibility is a detector for flying bird of prey. Lorenz (1939) and *Goethe* (1940) reported that young precoeial birds did not respond to flying geese but showed a flight response when a hawk or falcon passed overhead. Experiments by *Schleidt* (1961) showed that the result could have been due to habituation to the geese but not to hawks. Further work on nai mallard ducklings by *Green*, *Oreen* and *Ca* (1966) and by *Mueller* and *Parker* (1080), however, indicates that hawk and goose silhouet can be distinguished and do elicit different reactions. The reactions were not clear anti-predator reactions in either case. The lining of the perceptual recognition to the command which initiates such reactions, therefore, may require experience which the experimental subjects of these experiments did not have.

The general effects of experience of anti-predator behaviour have been examined in many experiments with young animals. If the occurrence of such behaviour depends, in part, on the gradual formation of a model of familiar surroundings, with which future events can be compared, individuals reared in different surroundings will form different models which may be formed at various rates. Experiments with rhesus monkeys, chimpanzees and domestic chicks indicate that the magnitude and direction of responses to novel objects or novel environmental changes is less for individuals with more complex rearing conditions. The large literature on the effect on later behaviour in a strange pen (open field) of handling rodent pups during infaucy, leads to the same general conclusion. It seems that the major effect of handling a rodent up is to change the amount of time that the mother spends licking and sniffing that pup. Perhaps as a consequence of this, handled pups show a less marked response in novel situations. The extent and type of maternal care also has effects on the responses of young primates confronted with possible danger. For example, *Ainsworth*. *Bell* and *Stayton* (1971) found that one-year-old human infants, whose mothers were inadequate in their responses to them explored less and fled to mother, mother more readily in strange and hence potentially dangerous situations.

3

Learning and Habituation

Learning—the modification of behaviour in response to experiences—is one of the most characteristic attributes of animals. The most important feature of learning is that it is adaptive. The animal, having learned, responds in ways that improve its survival and reproductive success. Different animals, even of the same species, learn different things if they are exposed to different environments. Learning is demonstrated when animals in two groups, one given an experience denied to the other, develop different behaviour patterns. Sometimes these experiments reveal that experiences do modify behaviour, but at other times they show that experience is unnecessary. For example, most young birds begin "practice-flying" before they leave their nest. They stand tall and flap their wings vigorously as if they were testing them for the first take off. But experiments have shown that birds can develop the ability to fly purely by maturation.

In one experiment, pigeons were reared in narrow tubes that prevented them from moving their wings. They could not undertake any form of practice flights. A control group of birds were allowed to practice each day. When the controls had "learned" to fly, the experimental birds were released from their tubes. Surprisingly, they flew just as well as the control birds. According to *Hilgard* (1956) learning occurs when the probability of certain behaviour patterns in specific stimulus situation has been changed as a result of previous encounters or other similar stimulus situations. *Hinde* (1970) has defined the learning as changes which can not be understood in terms of maturational growth process in the nervous system fatigue or sensory adaptations. According to *Wallace* (1973) learning is the adaptive changes in behaviours that result from the individual experience and probably associated with physical changes in the central nervous system. It may function with innate pattern or but may ultimately be separable from these patterns.

Learning, Adaptation and Natural Selection

It is also reasonable to suppose that behaviour and psychological processes such as learning and memory play a role in adaptation and may also be selected for. One way to visualize the role that behaviour may have played in evolution is to consider again some of the problems that an individual must solve in order to reproduce. Consider, for example, the necessity of food. If food tends to be unevenly distributed in the environment, it would be to the animal's advantage to be able to learn and remember routes to the best food sources, cues that predict the availability of food, the route back to the nest (if there is one), signs that predators might be in the vicinity of the food source and, if so routes to alternative food source. Also important would be the ability to learn and remember the distinguishing characteristics of toxic food substances. There is substantial evidence that learning is, in fact, important in food selection and foraging behaviour, and we will consider this in some detail in several places in this text.

For now it is important only to imagine that animals particularly efficient at these tasks would be more likely to produce offspring and hence pass on the ability to learn. Is learning in regard to food-related behaviour important in human culture? Just think of the substantial time allocated to such things as learning how to cook, recipe preparation, comparative food shopping, agriculture (requiring entire college curricula), the technology of fishing, and the dairy industry. Anti-predator behaviour is another problem faced by animals. Many animal species have evolved elaborate structural defenses such as camouflage colorations and morphology, and mimicry of toxic or dangerous animals. Some have developed evasive running behaviours; other have group defenses such as schooling (fish), herding (grazing animals), or mobbing (birds).

Learning also evolved to play a role in the safety of animals. For example, mobbing of a predator by birds not only serves to drive off the predator but also to acquaint the young with the nature of typical predators. For example, one species of monkey has different alarms calls for different types of predators and these calls must be learned by the young. It is interesting that the accumulated knowledge of human culture and technology is used for anti-predator behaviours in ways that mimic the structural and behavioural adaptations of other animals.

For example, in warfare (humans are essentially the only predators that humans have to fear) protective colouration is used. Ski troops wear white, jungle troops wear green, desert troops were khaki. There is even a "stealth" aircraft under development that is "invisible" to radar. Speed and evasive behaviours are used (in aircraft and ships, in particular). Repellent devices are used (weapons). All of these strategies employed by humans are based on learning in the sense of accumulated technology. It is not necessary to elaborate any further on how learning may have evolved (been selected for) in other survival problems that all animals face problems such as care of the young, habitat selection social interactions with conspecifics, etc.

The principal point is that survival problems can be solved in several ways by structural changes that adapt the animal to its environment; by innate behaviours based on complex reflexes or by behavioural changes based on the ability to learn. There is an important difference among these modes of adaptation. Adaptation resulting from selection of the ability to learn and behave flexibly provides an enormous advantage in that it contains the basis of further adaptations in the life-span of a single individual. An organism that has adapted solely on the basis of structural modification or innate behaviours faces difficulty if the environment changes; an organism that has adapted on the basis of learning may be able to learn new behaviours that will adapt it to the new environment. If this learning takes place, then such an individual will pass on its genes to future generations whereas the organism that has adapted solely on the basis of morphology or fixed behavioural mechanisms would be less likely to do so. Thus, the ability to learn would have been selected as an adaptive mechanism. Perhaps the flexibility of learning can be emphasized by describing some examples of complex behaviour that are surprisingly inflexible.

A species of wasp provides food for its offspring by paralyzing its insect prey and depositing it in an underground nest for the larvae to feed upon. In doing, this the wasps goes through a regular sequence of behaviour. When it first returns to the nest with its paralyzed quarry, it places the insect near the entrance and then goes down into the nest itself. The wasp soon returns, brings the insect down into the nest, and returns to the surface to close up the entrance. However, an interesting behaviour occurs if, while the wasp is on its initial inspection tour" of the nest, its prey is moved several inches away from where it was left at the edge of the nest. When the wasp now returns to the surface, it moves the prey back to the entrance and then goes down into the nest again—starting the sequence over. This apparently can go on and infinitum as long as the prey is moved each time the wasp enters the nest. It is as if the wasp were caught in some program "loop" that must be completed before it can go to the next stage of its provisioning sequence. The wasp, in this instance, demonstrates remarkably inflexible behaviour. Another example of inflexibility is the behaviour of the wasps Ammophila pubescens. This wasp may maintain many burrows, with larvae at different stages of development in each.

The wasp provides its young with food, giving each larva an amount appropriate for its size. Before setting off on the day's foraging the wasp visits all the burrows and them, eventually, returns with the proper amount of food for each burrow. If however, the larvae are switched after it has made its first visit, the wasp does not adjust its behaviour to the new occupants of the burrows. Instead, the amount of food placed in each burrow corresponds to the size of the offspring that has been there on its first visit—it does not correspond to the size of the current occupant. Again, the behaviour of the wasp is complicated and is a performance that requires a considerable memory. But the behaviour is also stereotyped and resistant to change based on changed conditions. In considering the evolution of learning, *Pulliam* and *Dunford* (1980) used the analogy of an investor and stockbroker. An investor could give his broker strict instructions to buy stock A, hold it for six months, then sell it and buy stock B, hold it for two years, and so on.

This strategy might work if the stock market were perfectly predictable and the investor could see the future. However, if the stock market were unpredictable and other, more favourable opportunities for investment presented themselves, the broker would be helpless without further instructions from the investor. This situation is analogous to the set of genes the has constructed an organism to respond to environmental contingencies in certain inflexible ways. If the environment remains the same as the one in which the organism evolved (was selected for), the programmed responses to contingencies should suffice to guarantee reproduction of the genetic material. But, if the environment changes, then the original program for survival (modes of adapting to the environment) might not be successful and, if so, the genes will be selected against and perhaps disappear from the population. In the example of wasp behaviour given above, it could be said that the wasp's genetic programming was not prepared for the environmental changes brought about by the experimenters. An investor could use a different strategy. He could give his money to a broker and instruct him to use his own experience and expertise with the stock market and invest the money as he saw fit, seeking the maximum return. If the broker were astute and knowledgeable, the investor would stand a better chance of doing well, in face of an unpredictable market, using this strategy rather than the inflexible instruction described previously. This situation is analogous to a set of genes that has constructed an organism with a nervous system that has the ability of learn and very its made of adaptation to match changing environmental contingencies.

If the nervous system learns well and applies its knowledge astutely, the chances of the genetic material being passed on to future generations is enhanced. Throughout this text we will see how learning builds on a base of innate reflexes, and allows each animal to face the challenge of survival with the inherited predispositions handed down through the eons from generations of ancestors, plus the unique knowledge that it may have acquired in its own lifetime. We will see how animals learn to ignore stimuli that seem unimportant, how they learn to anticipate changes in their environment, to prepare for these changes and, where possible to modify their environment and profit from the experience of being able to do so. However, before going on to a consideration of the simplest type of learning—habituation—we will first define learning more formally and specify some of the terms used in the experimental study of learning.

Learning Variables

The variables that is of interest in the experiment, the one that is manipulated by the experimenter, is termed the independent variable. The term independent is used because the experimenter decides whether or not to administer the variable and what levels of the variable to administer. In the two examples given previously, the different types of automobiles would be the independent variable in the collision-injury study, and the presence or absence of the drug would be the independent variable in the maze-learning study. The experimenter is interested in the effects of the independent variable on some other event. In order to determine these effects, the experimenter varies the independent variable while holding other variables constant. In order to assess the effect of the independent variable, some measure of the event of interest must be taken. The variable used as a measure is termed the dependent variable. It is termed dependent because changes in this measure dependent upon changes in the independent variable; it reflects the influence of the independent variable. In the examples above, some measure of damage to the mannequin might be the dependent variable in the automobile study, whereas the number of errors made in the maze might be the dependent variable in the drug study.

Generality of Learning

In the chapters the follow, we will be discussing particular areas of research in some detail. In each case, we will use experimental data to support the general conclusions that are reached. In most causes, cases, the experimental subject cited will be the rat because of the widespread use of this animal as a subject. But is it reasonable to base general conclusions on the behaviour of a specific organism? Would these same generalities apply to a sea slug, an insect a chimpanzee, a fish, or a college sophomore? In many cases, the differences in behaviour that would be observed by using another subject would be trivial.

The purpose of this text is to develop an understanding of the basic principles of learning, and it is the contention of the author that most of these principles are applicable across species. Although our major purpose is to develop general conclusions that apply to virtually all organism, we will also examine a variety of situations in which the nature of the organism clearly interacts with the nature of the task. Some organisms, for example, easily acquire an active running response to avoid an aversive situation, but have difficulty learning to be passive and sit still; other organisms show the opposite patterns. Similarly, it will be seen that many birds tend to use the colour of food as a cue in learning tasks in which rats respond to taste as a cue.

Even considering only one kind of organism—pigeons, for example—we will see that the precise way in which a pigeon pecks at a signal for a reward differs depending upon whether the reward is water or grain. Thus, although we will find many examples of generality in learning, we must also remember that each species evolved by adaptation to a particular niche in the environment and that both the structure and the brain of the organism must be somewhat specialized for the ecological niche to which it is adapted. For example, carnivores and herbivores have evolved quite different feeding behaviours. It would not be surprising to find that a cat (a carnivores and herbivores have evolved quite different feeding behaviours.

It would not be surprising to find that a cat (a carnivore) and guinea pig (a herbivore) behave somewhat differently in learning tasks that use a food reward. However, although differing in detail, the general principles of learning remain similar in such divergent species. Similarly, casts tend to be solitary hunters or hunt in very small groups, whereas canids (dogs, wolves) tend to hunt in large groups. Again, it would be surprising if the long evolutionary history that adapted these animals to such different social behaviours did not also lead to differences in brain structure that would influence some types of laboratory learning tasks. When an animal is taken out of its niche and brought into the laboratory, the experimenter must be careful to temper any conclusions about learning ability or learning principles with a considerable of how the learning task might relate to the animal's behaviour in its natural environment. Later in the text we will consider in some detail how the evolutionary histories of various species may interact with the degree and type of learning that they exhibit in the laboratory.

According to Wallace (1973) learning is the adaptive change in behaviour that result from the individual experience and probably associated with physical changes in the central nervous system. It may function with innate pattern but may ultimately be separable from these patterns. Many undergraduates who enroll in learning courses are interested in human earning and they see little value in the study of animal learning. There are, however, several reasons to pursue such investigations. One such reason is that there may be similarities in principles and/ or physiological processes underlying learning in humans and other animals. We will see many instances of this in subsequent chapters. For example, we will see how research on Pavlovian conditioning in animals has provided a way of understanding the learning of emotional responses in humans. It is quite likely that there are substantial similarities if not identities in the physiology, neurochemistry, and learning principles governing emotional behaviour across the animal spectrum.

Knowledge of these similarities helps in the understanding and treatment of phobic behaviour in humans, and it should soon help in the understanding of how emotional stress contributes to disease states such as hypertension and cancer. We shall also see how research in Pavlovian conditioning and food aversion learning in animals may have implications for the understanding of difficulties that cancer patients face when they undergo chemotherapy. Another example that may be cited in this regard is the widespread use and effectiveness of behaviour modification techniques in treating various behaviour problems.

The basic understanding of these techniques was derived from animal research. Many other examples of direct applications of animal learning to human behaviour will become apparent throughout the text. One such example is the relevance of the study of punishment and aversive control in animals for the understanding of the relative effectiveness of reward and punishment in shaping the behaviour of children. A second example is the relevance of the study of incentive relativity for the understanding of disappointment in humans. Even the use of language—very likely unique to humans—as an instrument to produce a particular outcome may follow some of the same principles as instrumental learning in animals, where they learn that certain behaviours will produce particular changes in the environment.

In other words, although the acquisition and many of the uses of language for congnitive processes may be unique to humans, the use of language as a tool in the service of notives and emotions may be similar to instrumental behaviour in lower animals. If these examples are relevant for the behaviour of humans, why not study them directly in humans? In some cases that is possible after the relevance of the animal research is perceived. In other cases it is not possible because of the importance of the principle of control. We have seen how extraneous variables must be controlled if the effectiveness of a particular variable on behaviour is to be understood.

In many instances it is not possible to control factors that might influence learning in humans—not possible because of practical reasons and/or ethical reasons. For example, if the effectiveness of a particular treatment on learning is to be investigated, how is it possible to ensure that two groups of humans come into the experiment with the same past experience: It is not possible. Also, human know when they are in an experiment and this knowledge itself may influence the outcome of an experiment. (There are somewhat related problems in animal research, but they are much easier to control). Thus, research that may have direct relevance to human behaviour is often best done with animals. A second reason for investigating learning in animals is that it may help us to understand what is unique about human learning. Although there are many examples of similarities between human and animal learning, it is clear that there must also be many differences. Other animals, for example, do not have differences.

Other animals, for example, do not have the language abilities that humans possesses—even the apparent rudimentary language structure seemingly demonstrated in chimpanzees is under question. There are probably many differences in cognitive abilities between lower animals and humans, and the extent of these differences will become clearer the more we know about animal learning. That is, the knowledge we have of animal learning may serve as background against which the unique aspects of human learning of human learning may be perceived more clearly.

Learning through the Conditioned Response

Animals with well-developed nervous systems can learn to behave in particular ways. Learned behaviour differs from innate behaviour in that the animal develops new responses to situations, and retains these responses for an extended period of time. Many experiments have been preformed to see if protozoans can learn. None of these experiments, however, has been able to show definitely that protozoans can learn. A few of these experiments are open to various interpretations, however, and some biologists might argue that some form of learning took place. We would be surprised if there were sharp line between animals that can learn and those that cannot learn. We might predict instead that there would be a spectrum of abilities from totally innate behaviour to the highest forms of reasoning. Sometimes, even in the vertebrates, it is difficult to know if learning actually has occurred. For instance, does a salamander learn to swim, or does it swim spontaneously when it reaches a certain age and stage of development?

In an experiment to answer this question, a group of salamanders were anesthetized at the age just before swimming movements normally begin. A control group was allowed to develop until they swam normally. When the experimental salamanders were allowed to recover from the anesthesia, they immediately swarm normally. These experiments therefore showed that swimming movements in the salamander are the result of normal development. Apparently learning has nothing to do with their swimming. One of the best-understood types of learning is the conditioned response, also called a conditioned reflex.

A conditions response involves the substitution of one stimulus for another. That is, assume stimulus A produces response B. If stimulus C can be made to produce response B, then one stimulus has substituted for another and a conditioned response has been formed. Experimenters have tried to develop conditioned responses in many types of animals. One of the most interesting experiments was performed on planarians. In the early 1950's, Robert Thompson and James McConnell, working at the University of Texas, developed a conditioned response in planarians. First, they shined a strong light on a planarian. Several seconds later, they administered a mild electric shock.

The planarian's normal response to the light was to stretch. Its normal response to the shock was to contract or turn its head. This sequence, in which the light was followed by the shock, was repeated about 100 times. Soon the response to the light alone was about the same as though the electric shock had also followed. This sequence is outlined. A "trained" planarian would show a shock response when the light was turned on 23 out of 25 times. It soon forgot its lesson if not retained periodically. In later experiments, trained planarians were cut in half. Each half was allowed to grow. The tail half grew a head; the head half grew a tail. The regenerated planarians were then conditioned to the light-shock response. It was found that both types of planarians, those regenerated from the head-half and these regenerated from the tail-half, learned quicker than planarians being conditioned for the first time.

Apparently some memory storage had occurred in planarians regenerated from tail pieces was well as in those regenerated from head pieces. You may want to refer back to review some of the regeneration experiments performed on planaria. These results suggested additional experiments to other biologists. Also taken into account were previous experiments on rabbits that had shown that RNA was probably associated with memory. Planarians were trained, cut in half, and

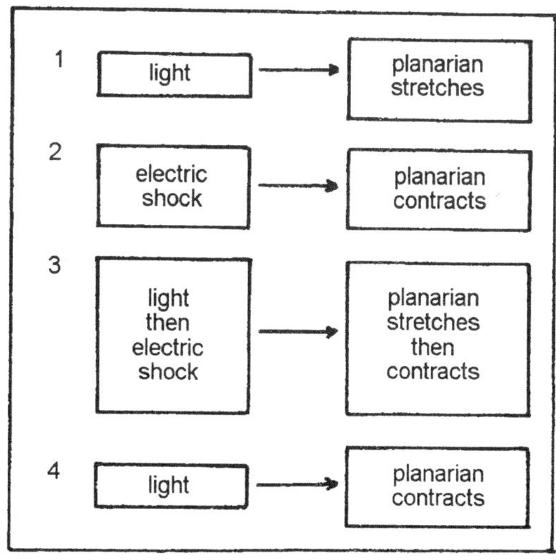

Fig. 3.1. The conditioned response in planarian.

allowed to regenerate as before. This time, however, the regeneration was performed in a solution ribonuclease (rye-both-New Kleeays), an enzyme which destroys RNA. When the regenerated planarians were retained, those regenerated from the tail-half learned about the same as a planarian which had never been trained, but those regenerated from the head-half learned at a significantly faster rate. It is not yet clear why there is this distinct difference in reaction between the two types. A second experiment with very significant results was designed to test the effect of a special type of cannibalism on memory storage. Trained planarians were chopped into small pieces. These pieces were then fed to a group of untrained planarians.

The control group consisted of planarians which were fed pieces of untrained planarians. The interesting result: the planarians that were fed bits of trained planarians learned better than the control group. These experiments on planarians have raised several questions: (1) Is the learning process a more common characteristic of life than previously imagined? (2) Are the nucleic acids, particularly RNA, the physical unit of memory storage? Such questions are currently being examined. Our interpretation of the trained planarian experiments may change when these questions are investigated further.

Imprinting are Experiments with Mazes

Most of you are probably convinced that baby birds do not have to learn to follow their mother. Yet, if a man squats down in front of a very young duckling that was hatched away from its mother, and if the man waddles along, quacking all the while, the duckling will follow the man. If the duckling is later given the opportunity to follow its own mother or other birds, it will still follow the quacking man. The duckling apparently learns to follow what it believes to be its mother. This form of learning is called imprinting. In investigating this form of learning further, it was found that a duckling will learn to follow a large coloured box with a ticking clock inside if the box is the first thing it observes. When the coloured box was pulled along on a wire, the duckling followed along, preferring the box to its own mother.

Imprinting is an unusual form of learning. These experiments certainly show that early attraction of a duckling to the mother bird is not instinctive—it is learned by imprinting because the mother bird is usually the first object that the duckling sees. For many years, a favourite way to investigate the learning ability of animals has been with the use of a maze. A maze is a series of passages in which an animal must choose between alternate paths. If an animal makes the right series of choices, he will be rewarded with food. If he makes the wrong series of turns, he may be punished, with a mild electric shock, for example. By being put through the maze again and again, an animal may learn to make the turns that result in reward rather than punishment. The simplest kind of maze is the T-maze where only one choice is involved.

Flatworms and earthworms can learn to make the "correct" choice of turns in this maze. Earthworms, for instance, are given the choice either of entering a dark, moist chamber or of receiving an electric shock. The earthworms took about 200 trials in the maze to learn to make the correct turn. After the earthworm had learned, it could make the correct turn in 90 per cent of the cases. If it had not learned, it would have been just likely to turn right as to turn left. A number of kinds of insects learn mazes quite rapidly, but even among insects learning ability differs. Ants are better learners than cockroaches, for example.

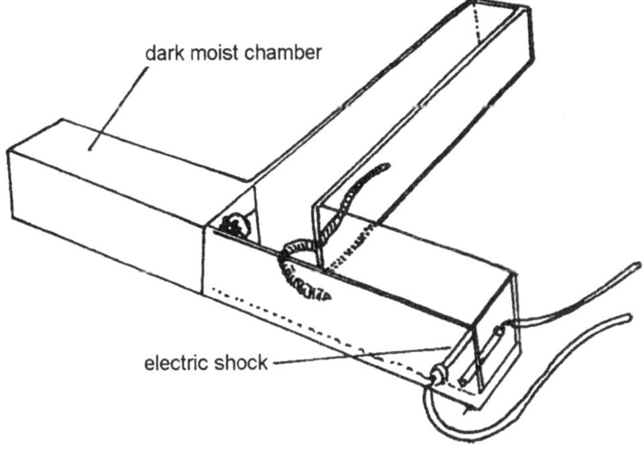

Fig. 3.2. An earthworm in a T-maze.

Reasoning is Characteristic of Man and Some Primates

The highest form of learned behaviour, and the one in which man excels, is reasoning. Reasoning can be used to solve complex problems without resorting to trial and error, the method used in maze learning. Reasoning ability can be studied with the "detour problem." In such a problem, and animal must follow an indirect path to get to a desired object, usually food. An animal must move away from the food before he can reach it. In experiments with the detour problem it is important to consider the following question: Did the animal solve the detour during its initial attempts or was trial and error involved? Of the many animals tested in this type of experiment, only monkeys and chimpanzees were able to solve the problem without extensive trial and error. Rats, dogs, and raccoons soon learned how to make the detour, but they failed to reason their way to a solution when first exposed to the problem.

Chimpanzees and monkeys also show the ability to solve more difficult problems. In a now classic case showing the reasoning ability of chimpanzees a banana was hung out of reach. By no amount of jumping could the chimp reach the banana. There were several boxes in the cage which, if stacked in the proper way, could provide access to the banana. How the chimp solved this problem. In any learning experiment, whether with humans or other primates, it is not always easy to decide how the problem was solved. Was it by reasoning or was it by the application of past experience? Or is reasoning actually an accumulate effect of experiences? The answers to these questions are at the heart of exciting work in the field of behaviour. How man and other animals learn is an important phase of scientific research.

Man is capable of much higher levels of reasoning than the other primates. He can learn to use symbols such as letters and numbers at a comparatively early age. These means of communication have made man's knowledge and civilization possible. That man's responses to letter and number symbols are forms of learned behaviour is made quite clear from the fact that no baby is born with the ability to read. He must be taught reading, writing, and arithmetic. These are basic skills for which only the ability to learn is innate. When he has learned these skills, he can then make further use of his extraordinary mental capabilities. The ability to reason, one of man's most cherished possessions, is also one of his least understood possessions.

Forms of Learning

Animal behaviourists have used many different procedures in studying learning. We shall next consider some of these. A description of the various forms of learning is analogous to the observation and description that provides the starting point for the study of all animal behaviour (*Tinbergen,* 1963). Many of the distinctions to be made among different "forms" of learning are based on differences in the procedures used in the study of learning. The fact that the procedures used differ need not necessarily imply that the processes underlying these different forms of learning are either similar or different. There may be one, two, or many kinds of processes underlying different forms of learning.

HABITUATION

Habituation may be defined as the decrease in probability or amplitude of a response, occurring when a stimulus which elicits that response is presented repeatedly. In an excellent discussion of the phenomenon of habituation, *Thompson* and *Spencer* (1966) listed nine characteristics of habituation that are manifested in most systems. We shall consider a few of these. If, after repeated stimulus presentation and habituation of the response, the stimulus is no longer presented, the response tends once again to be elicited by the stimulus. Within limits, the longer one withholds the stimulus the greater is the likelihood that the response originally elicited by the stimulus will again be elicited by it. This phenomenon is called *spontaneous recovery*. It may require a few minutes to several days, depending on the conditions. If a repeated series of habituation training trails and spontaneous recovery trails is given, the habituation occurs progressively more rapidly (*potentiation of habituation*).

Habituation of a response to a given stimulus shows generalization to other similar stimuli (*stimulus generalization*). Thus, if an animal habituates a response to a tone of a given frequency, it is likely that responses to tones of frequencies close to the training stimulus will also show habituation. The more dissimilar the test tone from the habituation tone, the less would stimulus generalization be apparent.

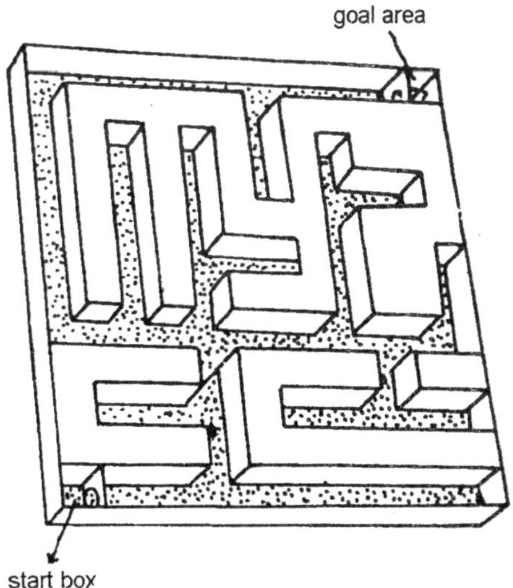

Fig. 3.3. A rat may learn a maze by trial-and-error processes, initially making many wrote turns and entering numerous clusde-sac.

Finally, presentation of another (usually very intense) stimulus results in a sudden recovery of the habituated response (*dishabituation*). In order to be considered true habituation, the response decrement must clearly be the result of changes in the central nervous system rather than of *adaptation* in the sensory system or fatigue in the effectors. When a puff of air is applied to the gills of a horseshoe crab, it responds with a movement of its telson (tail). *Lahue, Kokkinidis,* and *Corning* (1975) studied the characteristics of habituation of the telson reflex, monitoring the muscle movements electrophysiologically. They found a clear habituation of the telson reflex to repeated presentations of the air-puff stimulus. A brief squirt of saline from a syringe aimed at the site of habituation resulted in dishabituation. Initial levels of responsivity were normal after an interval of 1 hour without stimulus presentation (*spontaneous recovery*). Potentiation of habituation was observed after about 2 hours. No generalization of the habituation was observed when stimuli were applied to other regions (no stimulus generalization). Habituation of contraction responses of sea anemones to water-stream stimulation was observed by *Logan* (1975). *Ratner* and *Gilpin* (1974) studied the habituation and retention of habituation of the response to air puffs in both normal and decerebrate earthworms.

Examples of Habituation

Orienting Response

Habituation has been studied in complex responses of complex organisms, in simple responses of simplest organisms, and in numerous intermediate situations. We will present only a few examples. One of the more complex response patterns occurs in the case of the orienting response. The presentation of a novel stimulus of moderate intensity will elicit a variety of behavioural and physiological changes. For example, the organism will tend to orient toward the stimulus, there may be a decrease in its heart rate, a decrease in skin resistance, a decrease in respiration, changes in the construction of peripheral blood vessels, a change in muscle tone, and changes in the electrical activity of the cerebral cortex (Sokolov, 1963). If the stimulus is repeated a few times with no untoward consequences to the organism, then all of these responses will habituate, although perhaps at different rates. The habituation of one component of the orienting response—the change in skin conductance.

The presentation of a loud clicking sound on the first two trials elicits a large degree of skin conductance caused by activity of the sweat glands, which, in turn, reflects the activity of the sympathetic nervous system. As the loud click is repeated, the degree a change in skin conductance diminishes until eventually there is little response to the noise. The subjects in this experiment were humans separated into five different groups. Three of the groups were composed of highly experienced meditators using different techniques. These groups were instructed to engage in their standard meditating practice during presentation of the clicks. One of the other two groups was instructed to ignore the clicks (IGN) and the fifth group was instructed to attend to the clicks (ATT). Rate of habituation was approximately equal in all groups; apparently an inconsequential stimulus is an inconsequential stimulus in any of a variety of states of consciousness. Interestingly, Leaton and Jordan (1978) showed that habituation in rats to a loud noise, accomplished while they were awake, transferred to the sleep state. That is, rats that had been previously habituated were less likely to be aroused from sleep by the loud noise.

Gill Withdrawal Response

For another example of habituation, we turn from humans and rats to the lowly snail, in particular the sea snail *Aplysia california*. This snail, illustrated in panel A Figure grows to an adult length of about 20 centimeters and has a number of characteristics that make it an interesting organism in which to study habituation. The first of these is that Aplysia has a readily elicitable response that habituates with repeated stimulation. Tactile stimulation of the mantle shelf of siphon area of the snail causes it to retract its gill—a defensive reflex. However, if the stimulation is repeated and there are no adverse consequences to the snail, the degree of retraction becomes less and less, that is, the retraction response habituates. A second feature that makes Aplysia an attractive organism to study is that its habituation shows many of the same functional relationships that have been found to apply to habituation in more complex organisms. Panel C of the Figure shows that the response reappears after a rest period in which no stimulation is applied. This recovery is sometimes referred to as spontaneous recovery. Note that as the stimulation is repeated after the rest period, the response is rehabituated.

At first glance, this effect of a rest interval leading to the recovery of the response might indicate that the habituation is really due to a fatigue process and process and is not the result of learning. However, the next panel in Figure illustrates the results of one of several procedures that may be used to show that habituation is not due to fatigue. This fourth panel (D) shows that habituation can be counteracted

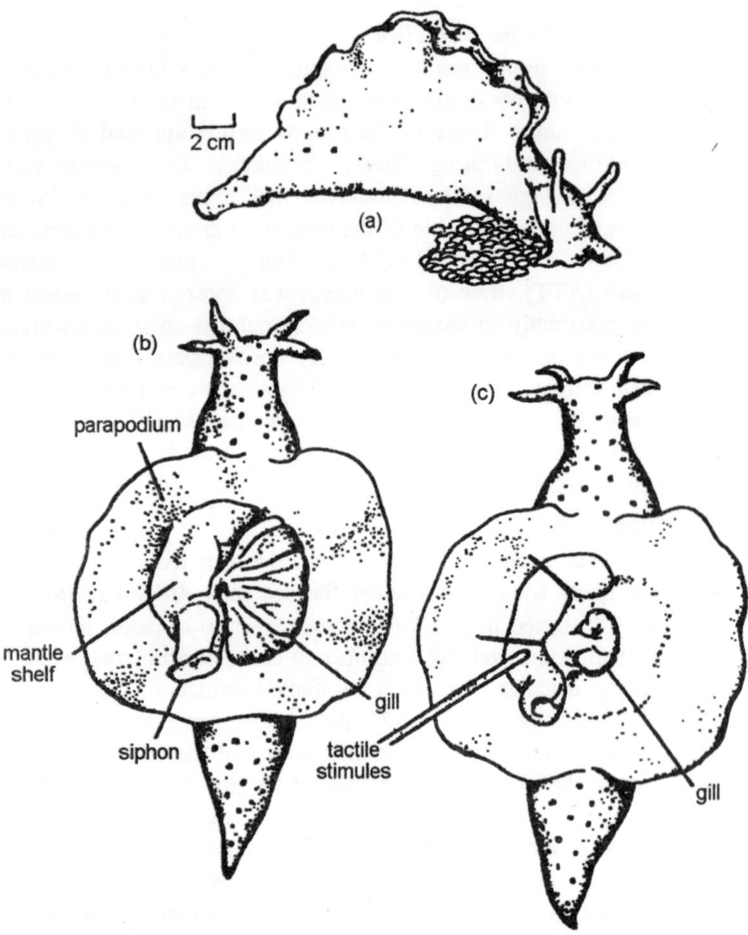

Fig. 3.4. The sea snail Aplysia californica. Tactile stimulation of the mantle shelf or siphon causes reflex retraction of the gill. This response habituates with repeated stimulation.

if the animal experiences another strong stimulus. For example, if Aplysia is touched somewhat roughly in the head and then the mantle shelf is stimulated, the previously habituated gill retraction response will reappear. This return of the response is referred to as dishabituation, and it indicates that the animal's muscle system is still capable of making the habituated response—it has not been lost because of fatigue. The next panel of Figure shows that dishabituation itself may be habituated. That is, the dishabituating stimulus has less and less effect in disrupting habituation as it is applied repeatedly.

The final four panels of the Figure illustrate two other relationships. They show that habituation is often more complete with less intense (weak) stimuli than with more intense (strong) stimuli, and that habituation may proceed faster if there is relatively little time between repeated presentations of the stimuli than if there is a long time interval between repeated presentations of the stimuli. We have shown the Aplysia has a response that may be habituated and that many aspects of this habituation are similar to habituation in more complex organisms. The reason that these characteristics are interesting is that the nervous system of Aplysia is relatively simple, indicating the possibility of our coming to an understanding of the neural basis of habituation. Whereas the human brain may have a trillion (or 1012) neurons, the nervous system of Aplysia has about 20,000 (or 2×10^4) neurons. Furthermore, there neurons are organized into nine separate groups (ganglia) each containing about 2,000 neurons. Many of these neurons are large and recognizable as particular cells occupying distinct

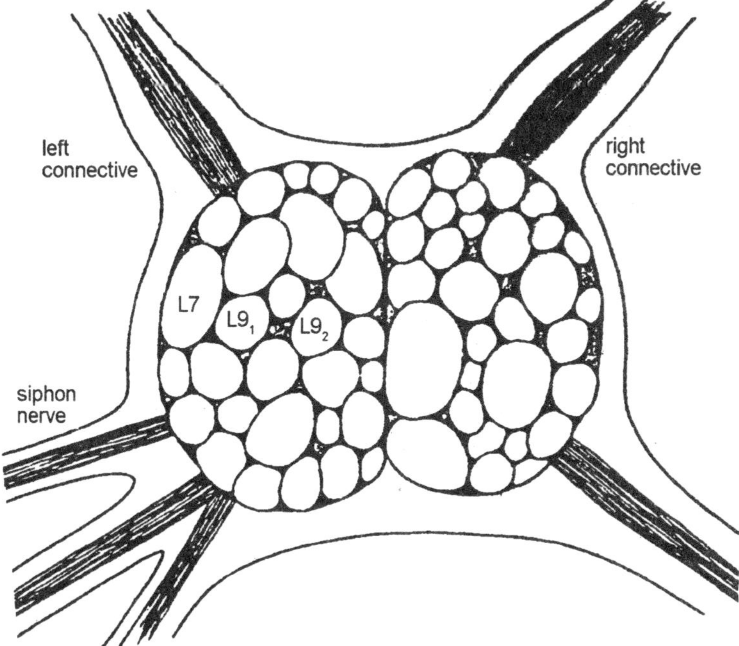

Fig. 3.5. Schematic drawing of nerve cells in the abdominal ganglion of Aplysia. Many of the neurons are large and are identifiable in regard to location and function in different animals. The neurons labeled L7 and L9 are involved in the gill withdrawal reflex.

locations and having distinct functions that are invariant in each individual. Some of this structure may be seen in the map of the abdominal ganglion of Aplysia presented in located in this abdominal ganglion (L7 is one of these motor neurons).

Neural Basis of Habituation in Aplysia

What happens at the neural level when a response becomes habituated? A long series of studies by Kandel and his colleagues has provided a substantial answer to this question in the case of the gill withdrawal response in Aplysia. A simplified version of their discoveries is presented in Figure. When the mantle shelf of siphon are touched, an impulse is initiated in sensory neurons. These sensory neurons influence. The first few times the sensory neuron is activated, there is substantial activation of the motor neurons and withdrawal of the gill. As the stimulus is repeated, there is less and less activation of the motor neuron. Does this mean that the motor neuron is becoming fatigued? No—because direct stimulation of the motor neuron shows that it is still capable of producing the complete gill withdrawal response. Is there less activity in the sensory neuron with repeated stimulation? No-Kandel's work has shown that the response in the sensory neuron is essentially undiminished by repeated stimulation. It seems that the change that underlies habituation occurs in the region of the synapse between the two cells. Kandel has been able to show that this change involves the release of less and less transmitter

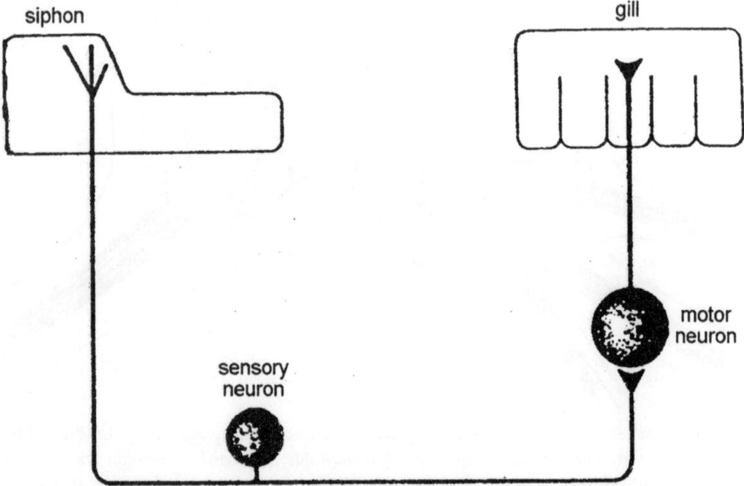

Fig. 3.6. Simplified "wiring diagram" of the gill withdrawal response in Aplysia.

substance by the sensory nerve as stimulation is repeated. With less transmitter substance released, there is less stimulation of the motor nerve and less of a gill withdrawal response. Is less transmitter substance being released because the supply is exhausted by repeated stimulation?

No—the phenomenon of dishabituation indicates that this is not likely to be the case and, in addition, Kandel has shown that the release of less transmitter substance is an active process that results from other changes in the cellular chemistry of the sensory neuron. Unfortunately, we will not be able to delve further into this cellular chemistry in this text. We will have to be satisfied with the knowledge that habituation of the gill withdrawal response in Aplysia has been traced to neurochemical processes taking place in individual cells. This knowledge will help greatly in the understanding of the neural bases of learning and memory in higher organisms. Before leaving the topic of habituation, however, we should indicate that even this simple form of learning is not as simple as we have presented it thus far.

One type of complexity may be seen in the field studies of habituation in sparrows by *Petrinovich* and *Patterson*. These investigators recorded the territorial song of a male white-crowned sparrow and played it back to a different sparrow pair through a speaker positioned within the territory of the pair. Various responses were recorded from the pair—for example, how often the male sang, how often it rose in flight, how closely it approached the speaker how often it attacked the speaker, how often the female took to flight etc. In addition, the songs were played back to sparrow pairs in different brooding conditions, that is, some were brooding eggs, some hatchlings, and some fledglings. The results of these studies indicated that habituation generally occurred, but the degree of habituation depended upon the brooding condition, the nature of the response, and the manner in which the song was presented. The point is that all responses may not habituate at equal rates and that the rate at which a given response habituates may depend on motivational states (here represented by brooding conditions) of the animal.

Different Processes

A second complexity is that all response changes that superficially appear to reflect the same process of habituation may not, in fact, reflect the operation of the same mechanism. For example, if rats are exposed to a loud click it will elicit a startle response that may be measured in terms of how much they jump. As the click is repeated, it produces less and less of a startle response. In other words, the startle response habituates. If a rat is exposed to a maze it has never experienced before, it will engage in a great deal of activity "patrolling" the maze. As the animal is given repeated exposure to the maze, its tendency to enter the various branches declines, that is, it seems to habituate.

Although behaviourally there would seem to be the same process occurring in these two situations–decline of a response with repeated exposure to the stimulus–there is evidence that two different brain mechanisms may be responsible for the two declines. Leaton has found that damage to the hippocampus, a part of the brain implicated in memory processes in humans, prevents the decline in exploration seen with repeated exposure to the maze, but such damage has no effect on habituation of the startle response. That is, the startle response habituates in about the same way in brain-damaged rats and in normal rats, but the rate of decline in maze exploration is quite different in the two groups. Thus, there may be a number of different brain mechanisms controlling different kinds of habituation.

In fact, it has been argued that maze explora-relations and the location of objects in space—and when the map is completed, exploration ceases *O'Keefe* and *Nadel* have also hypothesized that the hippocampus in necessary for the formation of these cognitive maps, and that without an intact hippocampus the animal cannot learn (or remember) spatial relations. Thus, exploration continues unabated. From this point of view, then, the apparently similar habituation curves of the startle response and maze activity would really represent two quite different processes–the learning of spatial relations in the case of the maze, and perhaps the learning that the loud click is unimportant in the as of startle.

Short and Long-Term Habituation

As a final complexity, we shall mention the possibility that there are two basic types of habituation: short-term and long-term. One way of demonstrating the difference in them is to conduct a relatively brief series of trails and them is to conduct a relatively brief series of trails and then check for recovery from habituation. Recovery usually occurs fairly quickly and completely. When, however, a longer series of trails is given, there is less recovery, and habituation is maintained over a longer time period. Another way of demonstrating this is to present the stimulus to be habituated at two different interstimulus intervals (ISI) to two different groups of animals. For example, one group could be presented a stimulus every two seconds, whereas a second group could be presented a stimulus every 16 seconds. When this is done, the short ISI group.

When however, these same two groups are tested some time after the initial habituation session, then the long ISI group might be likely to show less recovery from habituation than the short ISI group. It is possible that these two types of habituation may be related to difference in short-term and long-term memory in humans. Given that there are two types of habituation, it is necessary to postulate that there are two different neural processes responsible for them?

Some research by Kandel (1979a) indicates that the answer to this question is "yes and no." That is, in both short-term and long-term habituation, the sensory neuron of Aplysia becomes less effective in stimulating the motor neuron and the basis of this effect is that there is less transmitter substance released at the synapse. Thus, the same basic mechanism seems to be involved in both types of habituation. However, there must also be a difference since synaptic activity is depressed for longer time periods in long-term than in short-term habituation. Kandel speculates that there may be some structural changes in the presynaptic terminal of the sensory neuron that result from the extended habituation training. These structural changes then contribute to the longer-lasting effects of habituation. However, the effects of these structural changes are reversible, as we shall see when we discuss our next topic sensitization.

SENSITIZATION

Incremental Sensitization and Dishabituation

Sensitization refers to the augmentation of a pre-existing response by a strong, usually noxious, stimulus. There are two somewhat different ways in which the term sensitization is used. The first use is in referring to the enhancement of a response elicited by a stimulus with repeated presentations of that stimulus. We shall refer to this as incremental sensitization. It is clear from the definition the incremental sensitization describes a behavioural effect that is the opposite of habituation. A period of sensitization of ten precedes habituation. That is, the behavioural effects of the second, third or forth presentation of a stimulus way be greater than the effects produced by the first presentation. Subsequent presentations may, however, lead to less and less of a behavioural effect. Habituation, therefore, may follow a period of incremental sensitization. The first panel shows an initial marked increase in the frequency with which a male three-spined stickleback fish (*Gasterosteus aculeatus*) bites another male introduced into its territory. After the fourth minute of intrusion, habituation of the attacks sets in. The second panel shows a similar process in the number of song sung by a male white-crowned sparrow when the song of another male is played in its territory. In this case, the frequency of singing did not fall below the initial level during the course of the eight trails in which the "intruder's" song was played.

The third panel shows changes in skin conductance in humans as a function of the intensity of sounds presented to them. Remember from our earlier discussion that skin conductance changes are one component of the orienting response. Figure shows clearly that the degree of sensitization depends upon the intensity of the stimulus presented. When loud sounds are presented, there is a period of incremental sensitization that precedes habituation. A second way in which sensitization is often seen in behaviour occurs when a stimulus, usually intense or noxious, is presented just prior to the presentation of an already habituated stimulus. In this use, the term sensitization is synonymous with dishabituation. It is as if the intense stimulus primes the animal (sensitizes it) to respond to any change in the environment. By thus priming the animal, the new stimulus undoes the effects of the previous habituation period.

It is easy to imagine the potential adaptive value of the sensitization-as-dishabituation process. Once an animal has habituated to a stimulus, it is in a vulnerable state should that stimulus suddenly become dangerous. The occurrence of dishabituation following the experience of another stimulus that is intense or noxious may serve to reduce this potential vulnerability. That is, the conditions under which habituation occurred may no longer be the same, and it may be safest for the animal if its nervous system is constructed so as to return to initial conditions (dishabituation) when the environment is changed.

Dual Process Theory

What accounts for functions where responsivity tends at first to increase before declining? *Richared* Thompson and his colleagues have postulated that there exist two independent process that are aroused by the presentation of a stimulus: habituation and sensitization. That is, each presentation of a stimulus has a tendency to sensitize an animal to future presentations of that stimulus so that such presentations will produce an enhanced response. At the same time, each presentation of a stimulus leads to an increment in habituation such that future presentations of the stimulus will tend to elicit a smaller response.

The actual behaviour that occurs represents the outcome of these opposed processes. The two panels show hind limb flexor responses elicited by shock in cats. The cats in these experiments were prepared so that shock influenced only sensory neurons in the spinal cord, and the response of the cat's leg was controlled only by motor neurons in the spinal cord. The brain was not involved.

In addition to the responses obtained in this experiment (represented by the solid lines), the Figures also shows the hypothesized course of the two processes of habituation and sensitization (represented by the dashed lines). The behavioural data is assumed to reflect the interaction of these two processes. The first panel of the Figure shows data and hypothetical processes when the intensity of the shock was low. In this case there was no evidence of incremental sensitization. The assumption is that at low intensities there is little sensitization and the habituation process predominates. The second panel shows the results and hypothetical processes when a more intense shock was used.

In this case there was substantial incremental sensitization and little evidence of habituation. The inference drawn from these and similar data is that there is always some balance between habituation, which predominates when stimulus intensity is low, and sensitization which predominates when stimulus intensity is high. Some evidence in support of this dual process interpretation of habituation and sensitization has been obtained at the physiological level. For example, Kandel and his colleagues have studied sensitization in Aplysia. The found that a noxious stimulus applied to the head of Aplysia will greatly enhance a previously habituated gill withdrawal reflex. They have also found that this enhancement is brought about because of an increase in the amount of transmitter substance released by the sensory neuron. Recall that in our earlier discussion of habituation in Aplysia we stated that habituation was due to decline in transmitter release by the sensory neuron. Is sensitization of simple reversal of this process?

Kandel's work indicates that there is more to it than that. When the head is stimulated by the noxious stimulus, a neural system different from the one shown in Figure activated. This neural system, represented in simplified form in Figure, makes contact with the presynaptic terminals of the sensory neuron that arises from the mantle shelf area. When the head area is stimulated, this additional neuron system causes the sensory neuron to release more transmitter substance, thereby enhancing the degree of gill withdrawal. Activity in the neural system arising from the head ganglion serves principally to modulate activity

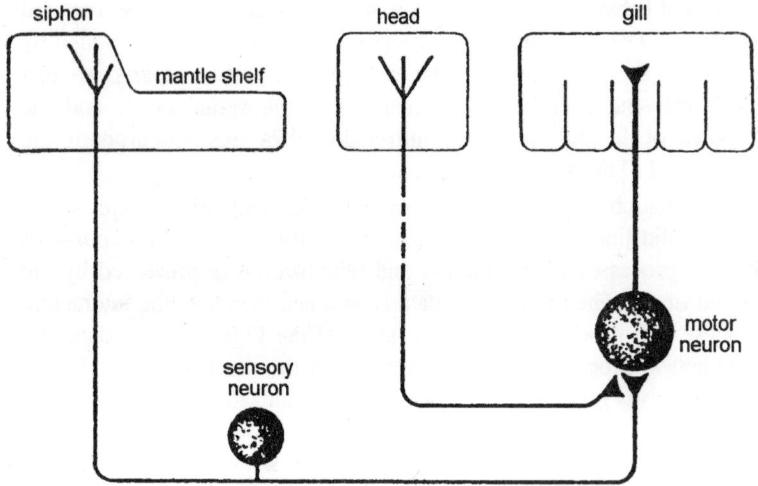

Fig. 3.7. Simplified diagram of dishabituation circuit in Aplysia.

in the sensory neuron. That is, it affects the amount of transmitter substance released by the sensory neuron only when the mantle shelf is stimulated. Stimulation of the head without subsequent stimulation of the mantle shelf has relatively little effect on transmitter release and gill withdrawal activity. Thus, the evidence adduced by Kandel and his colleagues agrees with the model proposed by Thompson and his colleagues—sensitization and habituation represent two separate and distinct processes.

Other research supports this conclusion. For example, research by Kandel's group has indicated that the chemicals serving as transmitters are different in the two systems, at least in the case of Aplysia. Whereas acetylcholine is the transmitter released by the sensory neuron, serotonin is released by the neuron from the head ganglion. Also, Thompson and his colleagues have found evidence that there are two types of neurons in the cat spinal cord: one type that shows only habituation and another type that shows sensitization. Thus, the evidence seems sufficient to conclude that the decrements in behaviour that we see as habituation, and the increments in behaviour that we see as sensitization, represent two distinct processes served by different neural systems.

In general terms, we could think of habituation as representing a process of learning that a particular stimulus is not important—it has no consequences for the organism. However, if an aversive event occurs, then a motivation system is activated which enhances reactivity to environmental changes, even those that have been previously habituated. Perhaps one more example will help to clarify habituation and sensitization. Imagine that you make it a regular practice to walk your dog at night in the dark, wooded back yard. There are a lot of sounds in the might—tree branches rubbing together, rabbits scurrying etc. However, both you and your dog pay no attention to them—you have habituated to these noises. Then one dark night your dog suddenly starts barking and growling, but you see nothing. The next, previously inconsequential, sound that you hear is likely to elicit some attention and perhaps some behaviour. In this example, the dog's barking served as a sensitizing stimulus, activating an arousing or alerting motivational system, which in turn dishabituated (sensitized) previously habituated behaviours.

Classical Conditioning

One very simple form of learning of the latter type is *classical conditioning*, so called because it was discovered by *Pavlov*, the father of conditioning. In his classical experiment, *Pavlov* lightly restrained a dog in a harness and repeatedly blew meat powder into its mouth and recorded accurately the amount it salivated. Then he associated the sound of a bell with the meat powder and repeated this procedure many, many times at successive intervals. The bell, of course, did not at first elicit salivation, but after repeated pairings with meat, it came to so. In describing this experiment, *Pavlov* called the salivation to the bell a *conditioned reflex* (CR), the bell a *conditioned stimulus* (CS), the salivation to the meat an *unconditioned reflex* (UCR), and the meat itself an *unconditioned stimulus* (UCS). This same experiment

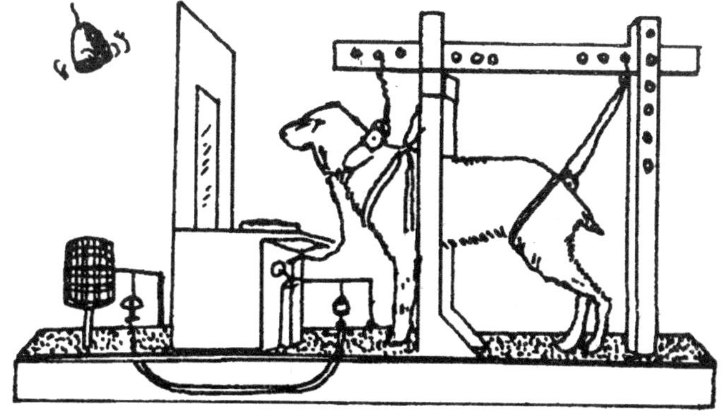

Fig. 3.8. Pavolov's testing apparatus. Ivan Pavlov discovered classical conditioning through his work on the salivary reflex in dogs.

has been repeated many times on different animals and with many different stimuli and responses. For example, the UCR may be a flexion of a leg in response to electric shock to the foot (UCS), and if this reflex is paired with the sound of a metronome (CS), that signal will eventually cause a leg flexion (CR).

Typically, the CR is very similar to the UCR, but it is never completely identical to it. Thus, the best way to describe classical conditioning is as *a process in which a previously neutral stimulus (CS=bell) is enabled to elicit a response (CR=salivation) that it never elicited before training.* In his experiments, *Pavlov* found that the time relation between the CS and the UCS was critical. If the UCS preceded the CS. There was very little, if any, conditioning, and if the UCS by more than about a second, conditioning became more and more difficult to establish instrumental conditioning in which the sensory stimuli involved become more numerous and more complex and in which the animal is given a greater and greater measure of choice in the stimuli it uses.

In general, as we go from the simple to the complex learning tasks, learning becomes more difficult and more easily disrupted by brain injury. With these points in mind, it will be interesting to compare the learning ability of different animals that represent different levels of the phylogenetic scale and thus different levels of development of the nervous system.

Operant Conditioning or Instrumental Learning

Operant conditioning or as it is sometimes called, *instrumental learning*, involves a wide variety of procedures. In each instance the animal learns to associate its behaviour with the consequences of that behaviour—that is, the sequence of events is dependent upon the behaviour of the animal. Usually some type of reward or punishment is involved. The task performed by the animal may be relatively simple, as in rat trained to press a lever in a skinner box. Pressing in this example is rewarded with food or water. Another example is when the animal has learned to run a maze to a goal box to receive the reinforcement. In nature a weasel may learn to associate the odor of mice with locating and catching a meal.

Mechanisms of Learning

It is easier to describe the learning abilities of animals than it is to understand the mechanisms are known. Completing the picture represents one of the greatest challenges to contemporary biology. There is good evidence that the mechanisms by which new information is

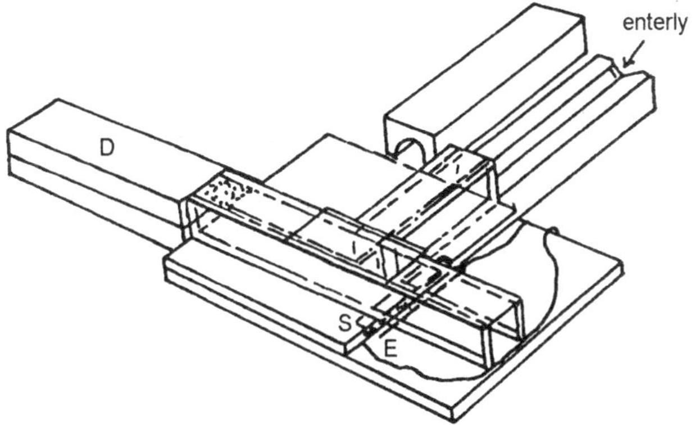

Fig. 3.9. T-maze used by Yerkes for training earthworms to turn to the right to enter a dark, moist chamber (D) and to avoid electric shock at E on the left.

first stored in the brain are different from mechanisms of longer-term storage. Brain concussions produce *amnesia* (forgetting) for events immediately prior to the injury, which is why people usually cannot recall what happened to them just before accidents. Short term memory may be based on continuously circulating nervous impulses. A rat or a hamster, just having learned a simple maze, such as a T-shaped runway in which the correct solution is to make a right turn at the junction does not remember having learned the maze if it is given an electric shock to its brain within 5 minute of its training. Slight amnesia effects are still detectable if the shock is given up to an hour after the learning experience.

In contrast, long-term memory is undisturbed by convulsions, electric shock, concussion, or deep anaesthesia. For many decades investigators have been trying to determine whether specific memories are stored a particular cells or groups of cells in the brain. Experiments have involved stimulating specific parts of the brain to determine what behaviour is evoked and maling lesions (cuts) in the brain to determine which memories are destroyed by particular lesions. Extensive work with rats reveals that limited regions of the brain are essential for retention of some memories but that particular cells or small groups of cells are not responsible for specific memories. The search for the sites of specific memories a more easily carried out with simpler animals that have fewer cells in their nervous systems. An especially favourable animal has been the marine gastropod mollusc *Aplysia*, commonly known as the sea hare.

As in most molluscs, the nervous systems of *Aplysia* consists of a number of distinct ganglia that innervate different body structures.

The cerebral ganglia innervate the anterior tentacles, mouth, eye and rhinophores. The buccal ganglia innervate the pharynx, salivary glands, esophagus, crop, gizzard and muscles that control the protraction and retraction of the odontophore. The pedal ganglia supply the foot, parapodia, head, caudal part of the penis, penis sheath, and penis retractor muscles. The abdominal ganglion controls a variety of respiratory, reproductive, circulatory and excretory functions and also controls movement of the external organs of the mantle. *Aplysia* is not a spectacular learner, but it does habituate to a number of stimuli.

Habituation of the defensive withdrawal of the siphon and gill under control of the abdominal ganglion. Electric shocks normally cause withdrawal of both organs, but if 10 stimuli are given at short (30-second intervals, the animal stops withdrawing its siphon and gills and does not resume doing so far several hours. Four such training sessions, each separated by approximately 1½ hours, produce habituation that lasts up to 3 weeks. Several neural changes are involved in this learning. One is a reduction in transmission of messages from the receptors in the siphon skin and the motor neurons that cause withdrawal. *Eric Kandel* and his associates at Columbia University have shown that just a few neurons decrease the amount of neurotransmitters they release. Conversely, when withdrawal is enhanced by following shocks with other negative reinforcement, a few neutrons release an amine that acts on the terminals of sensory neurons to increase the release of transmitters. Although withdrawal of both organs involves a number of muscles, changes in just a few neurons (the memory traces) are needed for the learning (the modified behaviour) to occur.

Phylogeny of Learning

We cannot retrace the steps in the evolution of learning, but a brief survey of the learning capacities and capabilities of organisms from various animal phyla should provide some insight into the possible generality of any laws of learning and a sampling of the differences in learning ability across the various phyla. Four cautions should be noted. First, there has been a disproportionate examination of learning across the various phyla. For more studies have been conducted on learning processes in invertebrates than in invertebrates, and even within the vertebrates much of the attention has focused on mammals. Second, there are valid ongoing disagreements among scientists regarding what constitute proper criteria for demonstrating various types of learning and concerning the evaluation and interpretation of learning and concerning the evolution and interpretation of research results. Third exploring the physiological and behavioural aspects of learning in many animals will require more refined techniques and objective, bias-free methods. Fourth, we measure performance and not a genetically determined ability or mechanisms. Lots of factors (notably, test design) can influence performance. Additional problems regarding comparative learning studies will be discussed in two subsequent sections of this chapter. With these difficulties in mind we can now survey the known information regarding learning in a number of animal phyla.

Protozoa and Coelenterata

Whether protozoa are capable of learning is still a debated question. It has been reported, for example, that *Amoeba* and *Paramecia* are capable of habituation to noxious sensory stimulus such as strong light or mechanical shock, for upon repeated stimulations their responses grow weaker and weaker until in some cases they become totally unresponsive. However, two criticisms of this work have been offered. One is that none of these experiments has satisfactorily demonstrated that such diminished responses lasted long enough to be anything more than adaptations to sensory stimuli. The other is simply that the noxious stimulation may have temporarily injured the organisms so that they were made capable of responding with each successive stimulation. To get around these objections, attempts have been made to demonstrate some form of associative conditioning or learning. But each claim has again been met with cogent criticisms.

One experiment will serve to illustrate the point. First a sterile platinum wire was lowered into the center of a dish of *Paramecia*. There was no special reaction to it. Next the wire was "baited" with bacteria, and the *Paramecia* responded by congregating around the wire, clinging to it and feeding. Then, after many such presentations of the "baited" wire, it was sterilized and dipped into the same spot, and the *Paramecia* congregated around it and clung to it. This was claimed to be a well-controlled demonstration of learning of a new response to the sterile wire. A simple control, however, demonstrated that the training was unnecessary. In this control experiment, bacteria were dropped into the dish and the *Paramecia* congregated and fed. Then the sterile platinum wire was lowered for the first time into the same spot, and the *Paramecia* clung to it.

Careful investigation showed that the congregation around the spot the wire contacted was due to the residue of bacteria bait. The increased clinging to the wire resulted from the increased acidity the bacteria contributed to the medium, since further controls showed that clinging

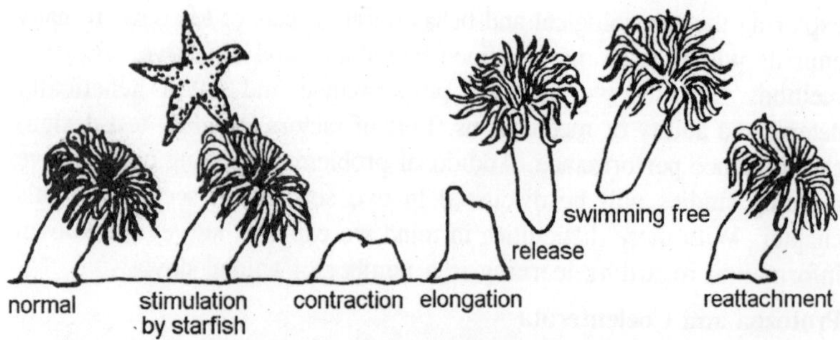

Fig. 3.10. Interaction between starfish and sea anemone.

in *Paramecia* was a function of acidity. For coelenterates there is also ample evidence for the habituation response, but little evidence for association learning. In the early years of this century investigators generally assumed that coelenterates exhibited only certain involuntary, stereotyped reflexes.

We now know that there are endogenous neural rhythms in many coelenterates and that, rather than being totally "passive," many coelenterates are capable of spontaneous active responses in the course of interacting with their environment. One possible demonstration of a form of association learning is based on the fact that sea anemones (genus *Stomphia*) respond to chemostimulation from certain starfish (e.g., *Dermasterias imbicata*) by stretching their bodies, detaching from the substrate, and "swimming" away. The animal soon lands and eventually reattaches to the substrate. The chemostimulation is paired with gentle pressure applied near the base of the anemone. With repeated trials the application of the pressure stimulus alone leads to some reduction in the "swimming" response; *Ross* labelled this as conditioned inhibition. However, the stimulus may have induced some related effect resulting in the anemone movement.

Platyhelminthes

Considerable publicity and controversy have surrounded the research conducted on learning to planarians. These flatworms are interesting because it is the platyhelminthes that a nervous system with bilateral symmetry and a primitive brain appears. In this case, a classical conditioning technique was used. When the worms were gliding along in a trough of water, a light was turned on, followed two seconds later by electric shock which caused the worms to contract longitudinally.

After 150 trails, the worms contracted to the light alone over 90 percent of the time. Following this, the worms were cut in half and allowed four week for regeneration. Both the regenerated head and tail sections showed a high degree of retention of what had been learned earlier. Even after these regenerated worms were cut and a second period of regeneration occurred, there was retention of the conditional response. Apparently, in this case, too, learning is not confined to the anterior portion of the nervous system, but has its effect throughout its extent.

Annelida

The first unequivocal evidence for learning is found at the level of the worms, where a bilaterally symmetrical, synaptic nervous system is already developed. Here we see clear instances of habituation and associative learning. In one experiment, earthworms were trained to go to one arm of a T-maze leading to a dark, moist chamber and to avoid the other arm which led to electric shock and irritating salt solution. In this simple learning of a position habit, it required about 200 trails on the average to reach a criterion of 90 percent correct responses. Interestingly enough, the worms were able to retain what they had learned after removal of the first five body segments containing the cephalic ganglion, and untrained worms were able to learn after removal of the head ganglion. Apparently, the neural changes involved in learning can take place in the ganglia of the lower body segments.

Mollusca

Among the molluscs, some evidence for the learning of a simple T-maze by snails has been demonstrated, but the experiment was not highly successful in that not all snails learned and those that did rarely reached high levels of consistency in their performance. Much better learning has been demonstrated in the octopus, which has a well developed eye and a rather elaborate brain. In these studies, for example, it was shown that the octopus could readily learn to discriminate the presence of a white card signifying electric shock. At first, the octopus was trained to come forward from its rocky nest to seize a crab that was lowered into the far end of the aquarium by a thread. Then on half the trails, a white card was lowered with the crab, and in these trials, the octopus was driven away from the crab by strong electric shock. After 12 trails, the octopus began to inhibit its approach to the card and by 24 trials consistently remained in its nest when the card was present and consistently came out to feed when the crab was presented alone. The octopus can also learn more difficult problems where two different cards were used, a small one associated with feeding and a large one associated with electric shock.

Also, discriminations of various shapes have been demonstrated in a similar manner. It is interesting, however, that the octopus was rather poor in problems requiring it to "detour" around a barrier. Thus, when a glass partition was lowered between the crab and an octopus, the octopus persisted in swimming straight into the glass and failed to swim around it through an open space.

Arthropoda

Habituation has been demonstrated in several arthropod species including horseshoe crabs (*Lahue* et. al. 1975), crayfish (*Krasne,* 1969), and the pupae of grain beetles (*Hollis,* 1963). A well-controlled study of classical conditioning of proboscis extension in blowflies was reported by *Nelson* (1971). The UCS was an application of a sugar solution to the mouthparts of the fly, with the CS being the application of water or a saline solution to the feet. Much attention has been devoted to the study of instrumental learning in arthropods. The studies of maze-learning in ants conducted by *Schneirla* (e.g., 1946) are classics in comparative psychology.

Effects of maze-learning experiences in larvae have been shown to survive through metamorphosis in adult grain beetles (*Borsellino, Pierantoni,* and *Schieti-Cavazza,* 1970). Conditioning of responses related to food getting in honeybees has been demonstrated by *Bermant* and *Gary* (1966) and *Wenner* and *Johnson* (1966). *Horridge* (1962, 1965)

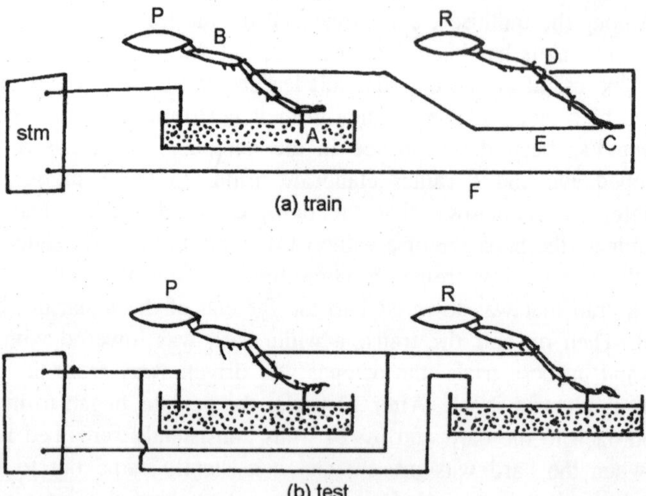

Fig. 3.11. Experimental arrangement for the conditioning and testing of insect avoidance learning.

studied the acquisition of avoidance responses in insects, with the without their heads, by using a rather clever procedure. In the training situation, the legs of two cockroaches are connected in series to an electrical stimulator. The electrical circuit is completed and both animals receive shock whenever animal P lowers its leg into a saline solution, thus completing the electrical circuit. With training, animal P comes to hold up its leg and thus avoid shock. Animal R is a "yoked control" that receives the identical number and temporal patterning of shocks, but they are delivered in such a way as to not be contingent on its behaviour.

Learning is demonstrated as a difference between animals P and R in a test situation when the animals are shocked independently for lowering their legs. Animal P experiences fewer shocks in the test situation than animal R. Both intact and headless locusts and cockroaches have displayed avoidance learning in this situation. Considerable difficulty has been encountered in demonstrating learning in fruits flies, a species that, in other respects, provides ideal subjects for behavioural studies.

Learning in Vertebrates

A number of learning studies have been conducted on species from most vertebrate classes. We know considerably more about the complex processes which underlie these phenomena for this phylum (*Masterton* et. al. 1967); this is particularly true for the mammals. Since the rat and Rhesus monkey have been used in a large number of these studies, an example for each species will provide some flavor of the research and how it is conducted. Rats can be trained in an operant conditioning apparatus to press a lever to receive food. They can also be trained in an apparatus with two levers (designated L for left and R for right) to press the levers alternately, LR or RL, for a food reward. Can rats be conditioned to press the levers in a LLRR or RRLL sequence, called double alternation? Investigators tested this question by conditioning rats first on the single lever task, either R or L, followed by the single alternation task, RL or LR. They then rewarded only double alternation performances.

The rats learned this task at a level exceeding chance expectations, and their performance improved over days. It was necessary, however, in some instances, to permit the rats to give extra responses to the first level before pressing he second lever. Thus LLLRR was rewarded the same as LLRR. When the investigators attempted to condition the rats to a sequence LLRRLLRR or RRLLRRLL (double alternation

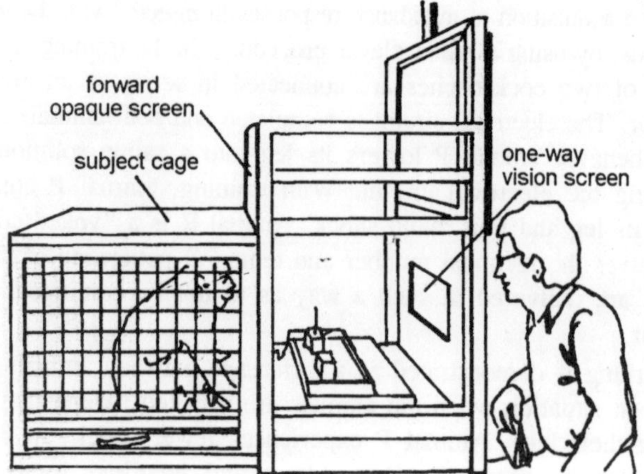

Fig. 3.12. Wisconsin general test apparatus. A early version of the Wisconsin general test apparatus (WGTA) used to test aspects of learning in primates and, with some modification of the apparatus cats and reccoons.

with a fixed ratio schedule of two repetitions), no rats could successfully perform at better than a chance level. The results are significant because of the new information provided regarding the capacity of the rate to associate a series of responses required to obtain the reward and because previous attempts to condition double-alternation tasks in rats had failed. One particular apparatus, the *Wisconsin General Test Apparatus* (WGTA) has been used in many of the studies of learning behaviour in macaques.

The basic procedure starts with the test tray out of reach but in full vie of the test animal. Food is placed in one food well, and then two identical objects are placed over the wells. After a prescribed delay the test tray is moved closer to the monkey, which then responds by lifting one object or the other. The four stages in the procedure are the baiting phase, covering phase, delay phase, and response phase. A trail ends when the monkey picks up one of the objects, uncovering either the correct food well, containing a reward, or the empty well. A number of variables can be investigated with this procedure, and other animals, such as cats and raccoons, have been tested with slight modifications of the apparatus. Among the variables that have been manipulated are length of the delay phase, the nature and size of the reward, the nature and the similarity or dissimilarity of the objects used to cover the food wells, and whether the animal is permitted to watch the test tray during the delay phase or, instead, has an opaque screen lowered in front of the tray. Among the conclusions are these

: Rhesus monkeys are capable of learning basic discriminations in this procedure with delays of up to 30 seconds or more, more food reward leads to performance, and imposition of the opaque screen during the delay phase increases error rates by up to 50 percent. One interesting and striking finding in these studies is that the behaviours and performances of individual monkeys differ markedly.

In general, monkeys that exhibit hyperactivity in the test situation and those that are more easily distracted during the delay phase exhibit lower levels of performance. Clearly, in studies of learning behaviour we must consider the significance of individual differences in performance, regardless of the species being tested, or the task being performed.

Neural Mechanisms of Learning

If we were to attribute in improvement of learning ability to one thing in the phylogenetic series, it would be the evolution of the central nervous system. It is an article of faith that learning represents some change in the central nervous system, and that memory is the preservation of that change. From this starting point, many investigators have sought an answer to two major questions: (1) where does learning take place in the nervous system, and (2) what is the nature of the change? We shall take up each of these questions separately although it is obvious that they are interrelated. The question of the *locus of learning* has been approached mainly by the technique of experimentally destroying parts of the nervous system.

Most investigators have dealt with the cerebral cortex, since the earliest theories held that it was in this newly evolved part of the brain that mammalian learning occurs. *Pavlov* believed that the cortex is essential for conditioning, but studies have shown that simple conditioning is possible in the dog after its cortex has been removed. Such a decorticate dog often give emotional and generalized responses and is greatly deficient in sensory capacity, but it can be successfully trained in the classical conditioning technique using shock as the UCS. Since total decortication grossly impairs the animal, many investigations of the cortex have involved the destruction of selected parts of the cortex. Thus, in his experiments on rats, *Lashley* just the visual area in the back of the cortex and tested the animals for visual learning and retention. He used a simultaneous discrimination situation in the jumping stand where the animal had to choose the correct one of two doors containing visual stimuli. When he used black and white doors, rats without the visual cortex could learn the discrimination almost normally. If they had learned to discriminate black from white before the lesion of the visual cortex, however, they lost the habit post-operatively and had to learn it all over again. When pattern discrimination was used involving a choice between a triangle and a circle, it turned out that the operated animals could never learn.

Apparently in this case, they lost the capacity for form or detail vision, whereas in the brightness discrimination, capacity was unimpaired and only memory was affected. Actually, however, further studies suggest that even in the brightness discrimination case, it was not memory that was affected, but rather it was a loss of sensory capacity needed to respond to the spatially separated black and white doors. To test this argument, dogs were confronted with a single, large, illuminated panel, shaped like a bowl so as to fill the entire visual field. Then they were conditioned to flex a leg every time the brightness of the field was changed. Here was a brightness discrimination not involving either spatial discrimination or the capacity to discriminate doors, and removal of the visual cortex had no effect on the animal's ability to retain it. So perhaps it was not a defect of memory that Lashley's brain lesions had caused.

Memory and learning ability also proved elusive in Lashely's maze experiments. Here he found that rats were affected in their ability to learn mazes, or retain them, in proportion to the size of the cortical lesions he made in their brains. In other words *Lashley* concluded that the cortex operates on a *mass action principle* in learning and memory so that the large the lesion, the poorer the ability. He also found that it did not mater where the lesion was in the cortex; a lesion of a given size had a given size had a given effect whether it was in the visual area in the back of the brain or the somatic sensory and motor areas in the front of the brain. From this finding, *Lashley* formulated his *principle of equiptentiality*, which says that all parts of the cortex are equal in their contribution to learning and memory.

Again, it seems that these experiments may be as much matter of sensory deficit following cortical lesions as they are the result of defects in learning and memory capacities. Maze learning, we know, is a matter of the rat learning to use many different sensory cues throughout the maze (vision, sound, touch, proprioreception, smell). The more these cues are experimentally eliminated from the animal's sue by destruction of sense organs or by removing stimuli, the worse its performance, regardless of which particular sensory cues are eliminated. Since the rat's cortex is primarily a sensory cortex, it is reasonable to believe that the large the cortical lesion, the more it will impair the use of sensory cues and, therefore, the poorer will be maze performance. Arguing somewhat against this interpretation is an ingenious experiment that *Lashley* performed. He taught blind rats a maze and then removed the visual cortex. Because they showed defects in the retention of the maze habit, he concluded that the visual cortex had nonvisual functions in learning and memory as well as visual functions. Many learning experiments of this type, involving various sensory capacities, have been done on different mammals, and the results have in general been the same. The cortex is not essential for learning or memory. The defects seen after cortical lesions are largely a matter of the sensory defects produced.

One exception to this get real statement is the recent work exploring the temporal cortex of primates. Work with monkeys shows that there are defects in learning touch discriminations following lesions of the posterior borders of the temporal lobe and defects in learning visual discriminations after lesions somewhat more anterior in the temporal lobe. Also, it has been found that human patients with bilateral temporal cortex damage seem to have defects of memory, especially recent memory. Finally, a most interesting related finding is the fact that electrical stimulation of the temporal cortex of fully awake epileptic patients evokes past memories in a vivid dream-like sequence. However, this may be possibly as much through arousing subcortical structures as through the effect on the cortex itself. Efforts to use the lesion method to explore the role of subcortical structures in learning have not been highly fruitful as yet. Not many such experiments have been done, and what information we have has been largely negative. Two recent studies, however, have provided hopeful leads.

In one, the experiments tried to condition brain-wave responses after making irritative lesions on one side of the brain by implanting aluminium cream. Of all the loci they investigated, placement of the irritative focus just below the temporal cortex in the amygdala and hippocampus was the most effective in impairing learning. In the second experiment monkeys were required to discriminate whether two patterns of tone or of light were the same or different, even though the two patterns might be separated in time by several seconds. In this case, surgical lesion of the region of the amygdala and hippocampus turned out to produce the most marked defects. Such operated monkeys could tell that the two patterns were the same only if one followed immediately after the other. It was as though the lesion made them unable to remember the first pattern over time, for they failed the test when the two patterns were separated by a few seconds. Another approach to the understanding of brain mechanisms underlying learning is through the use of electrical recording methods in which it is possible to trace changes in the electrical activity of many parts of the brain during learning. In these experiments, the animal has electrodes chronically implanted in its brain, and changes in pattern of electrical activity are noted as the animal is trained.

Striking thing here is that the changes take place in many parts of the brain, cortically and subcortically, within the sensory systems and outside of them as well. It may be that the electrical method is so sensitive in recording changes that go on in learning that it cannot separate the important from the unimportant parts of the brain for learning and memory. Or it may be that learning, or different facets of learning, occur in many places of the brain at once, and that therefore, no one part is completely essential for its formation or retention. When we come to the second question, concerning the nature of the change in learning, we find mostly theories and very little facts.

Various mechanisms have been suggested as being responsible for the establishment of new functional connections in the nervous system: (1) the growth of new have pathways, (2) anatomical swilling or sprouting of synaptic terminals, resulting in the facilitation of crossing Grain synapses, (3) a physiological increase in the ease of crossing synaptic connections already established but not functional at the start of training, and (4) biochemical changes such as alterations in the structural arrangement of protein molecules in nerve fibers. Of all these suggestions, one of the most intriguing has been the physiological concept of recurrent nerve circuit in which a loop or circle of connecting neurons is activated such that each neuron in the circle activates the next until the first one, having time to recover, is activated again. Such a loop could theoretically continue firing indefinitely and serve to add facilitation to any synapses in makes outside the loop and thus provides the basis for long-term memory.

At the present time, however, neither this nor any of the other theoretical suggestions have any direct evidence bearing on them. Some insight into the nature of memory mechanisms has been gained by direct, experimental examination of temporal characteristics of the memory process. In one study, rats were trained to run from one compartment to another to avoid an electric shock. They were given one trial a day, and after each trial, they received an electro-convulsive shock through the head. Different groups of rats received the convulsive shock at different times after the learning trail 20 sec, 1 min, 4 min, 15 min, 1 hr, 4 hr. If the shock came within an hour, there was virtually no learning, but if it came after four, hours, learning was essentially normal. Apparently, memory takes time to "set" in the brain, and this consolidation requires at least an hour, which suggests that memory is a two-part process, consisting of an early phase when memory is vulnerable to convulsive shock and a later phase when it is not. The same kind of conclusion turns up in two other rather different studies.

In one, the octopus was trained, as we mentioned earlier, to discriminate between a crab it could eat and a crab, accompanied by a white card, that it could not approach under penalty of electric shock. If the octopus vertical lobe, an associational region of the brain, was removed, then a curious thing happened. If the trials were spaced more than an hour apart, the animal could not retain enough from trial to trial to improve its performance in normal fashion. If the trials were within fifteen minutes of each other, however, it was able to learn easily. Apparently, lesion of the vertical lobe affected the "permanent" laying down of memories, but did not disturb their "temporary" establishment.

In man, lesions of the temporal cortex on both sides of the brain may result in a similar defect. These patients can learn something simple and retain it for about fifteen minutes to an hour, but after the time, they forget completely and may not even remember having learned. Yet their life-long memories are left undisturbed. Taken together with the animal studies, this finding suggests that memory is a two-part process: (1) an initial, vulnerable, perhaps physiological process lasting fifteen minutes to an hour, and (2) a later, invulnerable, perhaps anatomical process, providing the permanent basis for memory. Many mysteries remain in our quest for the physiological basic of simple learning. We know that learning is a property of is a property of at least all animals possessing a synaptic nervous system. The major question is whether the superior learning of mammals and especially primates is due to the development of superior neural mechanisms for learning and memory or to their greatly increased sensory and motor capacities or both.

The development of superior neural mechanisms must be an important factor, for the capacities for complex learning, problem solution, and reasoning emerge with the evolution of the central nervous system. We have reviewed, in this chapter, the basic facts of animal learning. We began with concept that learning represents an *enduring modification of behaviour* brought about by experience. Then we described various kinds of learning from the simple to the complex: habitation, classical conditioning, instrumental conditioning, and trial-and-error learning.

The essential modification in behaviour in all these cases in the development of some new response to a stimulus that nerve before elicited that response. As to the critical elements of the experience in learning, they appear to be very much the same in all these cases. Or put another way, these various instances of learning including human verbal learning, all seem to obey the same fundamental laws of learning: contiguity, repetition, reinforcement, and for the case of extinction or forgetting interference. When we took up the *phylogenetic development of learning*, we could not find clear-cut and reliable evidence for learning until the level of the worms. This is the point in phylogeny where the bilaterally symmetrical, synaptic nervous system first appears. With the cephalopods and arthropods, which have relatively large, concentrated ganglionic masses in the anterior regions of the nervous system, learning ability is much greater than in the worms.

Finally, with the development of the vertebrate brain, learning capacity develops even further, gradually reaching an asymptote among the simpler mammals. Despite this evidence that relates learning ability to the development of the central nervous system in phylogeny, it has not been easy to discover, with any degree of specificity, the neural basis of learning. The evidence from experimental brain lesions and, more particularly, from studies that record changes in the electrical activity of the brain during learning suggests that learning takes place in many places within the brain at once.

The nature of the neural change in learning has proven elusive, however, despite the fact that many attractive and plausible theories have been proposed. At present, we know that learning or, more particularly, the formation of memories is at least a two-part process. Initially, there is a temporary, perhaps physiological process lasting up to an hour in the mammalian nervous system. Following this and perhaps as a result of it, there a second, more permanent, perhaps anatomical change. Still a third mechanism may subserve the storage of long-standing memories, for it is possible to impair the permanent laying down of new memories by brain lesion without impairing either old, long-standing memories, or the temporary acquisition of new memories.

4

INSTINCT BEHAVIOUR

Almost all animals are composed of the same basic materials, the difference between species resulting from differences in the way in which these basic materials are put together. Although this regulatory function is poorly understood, it is known that genes control development by producing proteins that regulate the complex organization of embryological progress. How does an animal's behaviour become so well fitted to its normal environment? There are two basic ways. Firstly, it may be born with the right responses 'built in' to the nervous system as part of its inherited structure. Honey-bees inherit the ability to form wings and wing muscles for flight; they also inherit the tendency to fly towards flowers and seek nectar and pollen. Such responses are popularly called 'instinctive'—a term which has often been abused bur remains useful.

Instinctive behaviour evolves gradually, as do structural features, and natural selection modifies it to fit the environment in the best way. It forms a kind of' species memory' passed on from each generation to its offspring's. Alternatively an animal may have no inherited responsiveness with regard to a situation but instead have the ability to modify its behaviour in the light of experience. It learns which responses give the best results and changes its behaviour accordingly. Instinct and learning both ensure adaptive behaviour, the former by selection operating during the history of a species, the latter during the history of an individual. Stated in this way there is a clear dichotomy, which is realistic when actual examples are examined. But before we do this it is worth considering the importance of instinct and learning in a general way through the animal kingdom.

Conditioned Reflex

Watson studied learning in animals such as rats and was impressed by the flexibility of their behaviour and by how learning could lead them to behave in an adaptive manner. He believed that they built up connections in their brains as a result of perceiving associations in the outside world, so they would associate a particular place with food and run through a maze to get there. His views were partly stimulated by the work of *Ivan Pavlov*, the Russian physiologist, who had studied conditioned reflexes in dogs. *Pavlov* showed how a new stimulus could be made to elicit a reflex as a result of the animal building up as association. In his most famous experiment, a dog was presented with food, a stimulus which normally makes dogs produce saliva in readiness to eat, and at the same time a bell was rung. This normally has no effect on salivation. After a number of such presentations, the bell was suddenly sounded without any food being produced the dog salivated.

The animal had formed an association between the bell and the food so that both were now capable of producing the response. This sort of training is referred to as conditioning and the result, in this case salivation in response to a bell, as a conditioned reflex. The main stress in the theories of *Pavlov and Watson* was not on rewards : the dog did not need to eat the food to form the connection. Later ideas have suggested that a great deal of learning does rely on reward, the most notable exponent of this view being *B.F Skinner*. Like *Watson* before him. *Skinner* believer that the great majority of behaviour, with the exception of simple reflexes, results from experience. But he stresses the way in which the responses of animals come to be linked to particular stimuli as a result of reward. Animals tend to repeat rewarded actions and reward can thus be used to alter their behaviour.

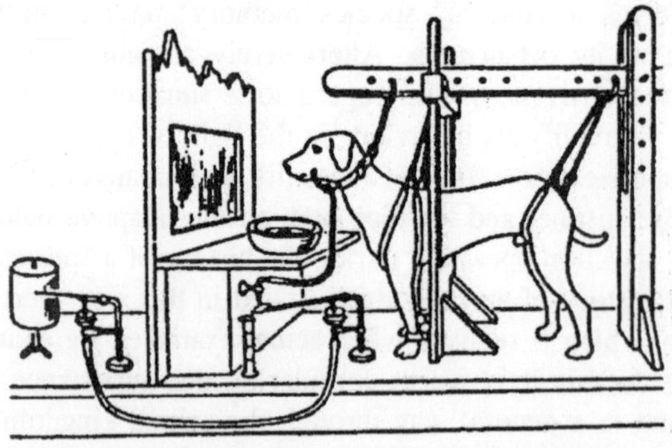

Fig. 4.1. Animal in a set-up for a Pavlovian study of conditioned responses.

The dove in the Skinner box referred to in the last chapter soon learns that pecking the right-hand key gives it water while the left-hand key yields food. Based on ideas such as these, elaborate theories of animal learning have been built up which stress the role of the environment and of experience, rather than genetics and inheritance, in the development of behaviour. At an extreme, such psychologists have suggested that learning in all animals is subject to the same rules and that animals do not have to have predisposition's about the sort of things they will learn but will build up any associations with equal ease provided that their sensory and motor systems allow them to do so.

Obviously one would not attempt to train a sparrow to fly in the dark or a tortoise to walk on its hind legs. The following famous assertion of *Watson's* illustrates the fervour of his belief that the environment rather than heredity was the major determinant of behaviour : 'Give me a dozen healthy infants, well-formed, and my own specified world to bring them up in and I'll guarantee to take any one at random and train him to become any type of specialist I might suggest doctor, lawyer, artist, merchant-chief and yes, even beggar man and thief, regardless of his ancestors.' He went on to admit that he was over stating the case, but insisted he was doing so as a counter to the sweeping claims of his opponents.

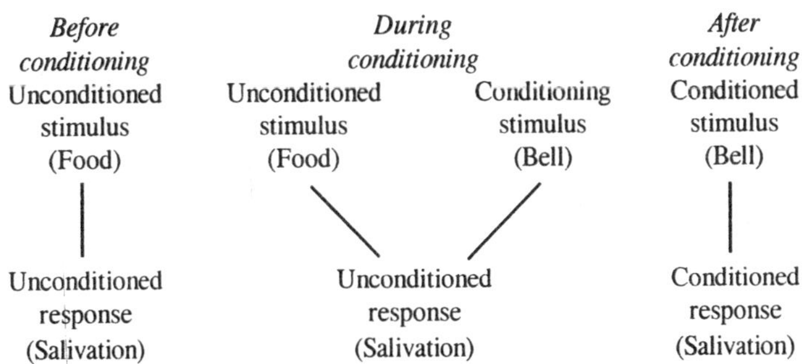

Fig. 4.2. Conditioning involves the formation of new connections.

Both views such as this and those of the ethnologists could not be right, but to some extent they existed because their adherents were studying different things. Psychologists were interested in learning and in behaviour patterns, like the lever pressing of a rat in a Skinner box, which could be modified by it so leading an animal to behave differently from other members of its species. By contrast, the ethnologists were interested in more fixed behaviour, such as courtship displays, which were common to all members of a species. Why did they consider such a behaviour to be 'innate' and how does the evidence they used hold up to detailed examination?

Ethnologists and Instinct

In 1828 a boy of about 16 was found in the market-place at *Nuremberg* in Germany. He could not speak but his behaviour was like that of a child, so he was labelled the *wild boy*. His name was *Kaspar Hauser* and, when he was able to communicate, he explained that he had been brought up entirely in isolation by a man who had kept him and cared for him all by himself in a hole. *Kaspar Hauser* may well have been a fraud, but he has given his name to the deprivation experiments which have often been used by ethnologists in an attempt to understand whether or not the behaviour they were studying was innate.

The idea was to deprive an animal of relevant aspects of experience and, if the behaviour appeared despite this, it could be regarded as inherited whereas, if it did not appear or developed abnormally, it could be assumed that the missing experience was important. There are many striking examples of behaviour appearing apparently normally in animals whose experience is distinctly abnormal. A case in point was shown by experiments carried out by *Janet Kear* on ducklings which she has reared by hand from the egg. Most duck species nest on the ground, but some of them, such as the wood duck, do so in holes far up in trees from which the fledglings have to leap to the ground beneath. Fortunately they are light and fluffy when they do so and, as a result, they bourse and run off rather than suffering multiple fractures.

Kear tested chicks of various species on a visual cliff such in this apparatus the animal is places in the centre and can move either to one side where there is a drop looking like a cliff beneath the glass or to the other where the floor is immediately under the glass so that it looks shallow. The behaviour of the ducklings was appropriate to their normal nesting place. Tree-nesters did not avoid the deep side of the cliff but, if they moved that way, they would leap as if casting themselves into space. On the other hand, the ground-nesters tended to move to the shallow side rather that the deep one, suggesting that they avoided heights. Interestingly, if they did move to the deep side their behaviour was quite different they pushed off with both feet as they would when moving out from the edge of a pond! Being hand-reared, these young birds had and no opportunity to learn the actions they showed from others or from earlier experience with ponds or with cliffs.

Many ethologists would therefore have used this evidence to argue that the behaviour must be innate because it develops despite deprivation of opportunities for learning, and the main motivation for carrying out such experiments has often been to discover whether behaviour is *innate* or *learnt*. Sometimes, as with the ducklings, behaviour develops normally even though the animal is reared in a very impoverished environment. Another good example here is the hoarding behaviour of squirrels whereby, even in captivity, they will bury nuts underground to form stores which they eat later when food is scarce.

If such an animal is raised on a liquid diet, so that it never experiences nuts, with masses of this food available the whole time, so that it never needs to hoard, and on a bare floor so that digging is impossible, the first time it encounters nuts and earth it still digs a hole and buries them. By contrast with these experiments, others have shown behaviour patterns to be radically altered unless particular experiences are available. If a young dog is reared in total isolation from all others, in a chamber in which it can be fed and cared for without contact with even its human caretakers, it turns into a strange animal, apparently careless of its own welfare. It will put its paw in a fire and singe it, and it will repeatedly approach an object that gives it an electric shock. This is quite unlike a normal puppy of the same age, which hastily withdraws from such painful experiences. Furthermore, the normal puppy will yelp when hurt, whereas the one that had been isolated behaves as if it did not even feel the pain.

Clearly its deprivation has led to a drastic alteration in the way its behaviour developed. In another example, to which we will return later, a young male chaffinch reared out of earshot of all birds of its species has been found to develop a very simple and unstructured song quite unlike that of a normal adult. If he is deafened as well, so that he cannot even hear his own efforts at singing, the song he produces is even worse, bring little more than a screech. These *Kasper Hauser* experiments vary enormously in what they actually deny the animals. The isolated chaffinch cannot copy song from other birds, but he can practice singing; if deafened he can still practice, but cannot here the outcome.

The hoarding squirrel has had experience of neither nuts nor earth so its deprivation is more extreme: it cannot copy from others, it cannot learn for itself, nor can it practice digging in earth, though it can carry out the movements concerned on the bare cage floor. In many cases it has proved very difficult to deny animals all the experiences one might think likely to be relevant. Finches with no hay that they can use for nest building will carry seed, lettuce, faeces and feathers to their nest site. Some bizarre behaviour results when the feathers used are still attached to themselves or to their mates. A bird may have to walk rather than fly to its nest site because it is holding its own wing in it beak as nest material; after carefully placing the wing in a corner of the nest it will then fly down, pick the wing up again and struggle back to the nest with it!

Characteristics of Instinct and Learning

We must now examine more closely the characteristics of instinct and learning. There are two conspicuous features of instinctive behaviour which may seem, at first sight, to be unique. First, that it consists of rigid, stereotyped patterns of movement which are very similar in all individuals of a species; all differ-wasps of the same species build their nests in the same way, domestic cockerels all use the same series of movements when courting hens, and so on. Secondly, instinctive patterns can often be evoked most readily by very simple stimuli. When presented with a complex situation the animal responds to one part of it and virtually ignores the rest. A robin displays more aggression to a tuft red feathers from the breast of a rival male than to a complete bird which lacks only these feathers. However, striking, such characteristics are quite inadequate to distinguish instinctive from learnt behaviour.

The latter is often described rather vaguely as being 'more flexible', but in fact the patterns of movement involved may be just as stereotyped as those of instinct. Rates placed in a box where they must learn to press on a projecting lever on order to obtain a pellet of food will develop a particular manner of doing so. Some use always their left paw, others press with their chins, and individual rats tend to be conservative in the method they use. In his delightful book, *King Solomon's Ring, Lorenz* describes how water shrews learn the geography of their environment in amazing detail. If at one point on a trail they have to jump over a small log, this movement is learnt with such fixity that they continue to make the jump in precisely the same fashion long after the obstacle is removed.

There are plenty of other examples where animals have comparable difficulty in 'unlearning' something. The movements appear to become almost 'automatic' and may persist even of they are no longer effective.

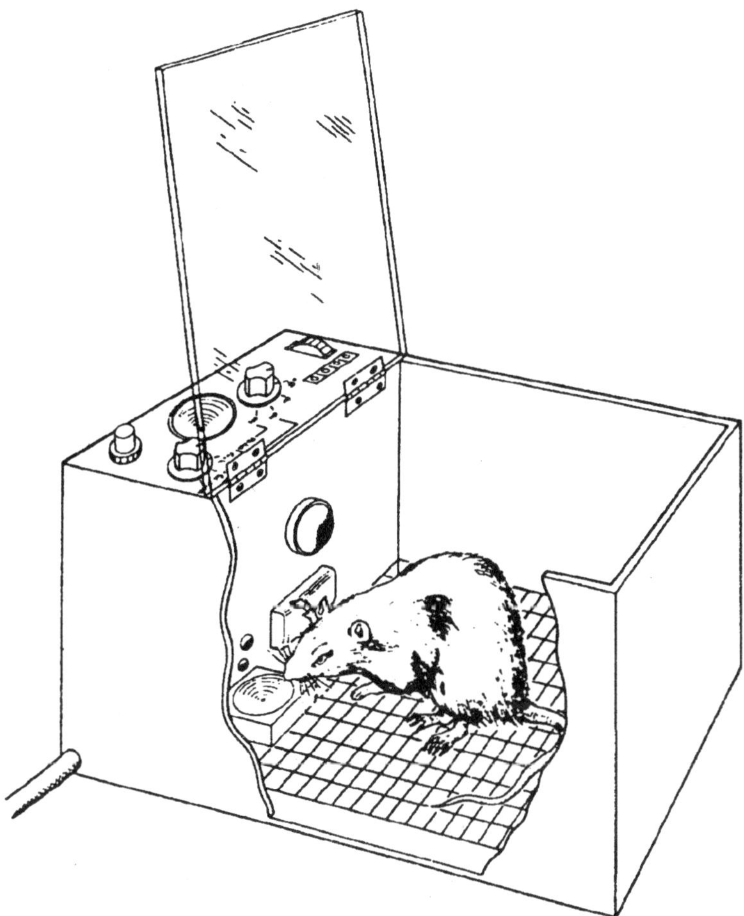

Fig. 4.3. Rat in a Skinner box pressing the lever which delivers a small pellet of food into the cup.

Learnt patterns nearly always involve responding to particular cues in the environment and, just as with instinctive behaviour, other features may be ignored. It is relatively easy to train animals to discriminate one key stimulus within a complex changing situation. If we are to make any critical distinction between instinctive and learnt behaviour; it must be based not upon their overt characteristics but upon their development within the individual. Rats learn to press a food lever and each does so in a stereotyped way, but this way is unique to the rat and varies between rats. All sexually receptive female rats show the same stereotyped acceptance posture when mounted by a male.

Furthermore, if we put a hungry rat into a box with a food lever, it may be hours before, by chance, it pushes against the lever. But if a virgin female rat is brought into receptive condition by a hormone injection, it assumes the acceptance posture after very little experience with a male. Here we have the basis for developmental criteria of instinctive behaviour and the commonest way of testing them has been the so-called 'isolation experiment'. Animals are kept individually out of contact with others from as early an age as possible. When mature, their responses to a variety of stimuli are tested and compared with those of animals reared normally. Rather few animals have been tested under really rigorous conditions but some fish and birds have been shown to perform various feeding, sexual and alarm patterns of behaviour quite normally after being reared in isolation.

The life histories of many insects are natural isolation experiments which demonstrate vividly that no practice or learning are involved in the behavioural development of the adults. However, easily isolation experiments enable us to eliminate the possibility that animals learn how to do something, they offer only a very restricted view of the factors that may be operating during development—learning and practice are only two from a wide range of possibilities. The point is that isolation can tell us only what factors are *not* important for the development of behaviour, it tells us nothing of what is involved and this may not be at all obvious. For example, young mallard ducklings hatched from an incubator respond preferentially to the call notes of mallard ducks over other related calls the first time that they hear them. They clearly show an instinctive preference for their own species call. However, Gottlieb 171-2 has shown that one factor which contributes to the development of this preference is the embryo duckling's ability to here the calls—quite unlike those of the mother—that itself and other ducklings make whilst still inside the eggshell. Nobody would doubt that the development of many behaviour patterns must be under genetic control and result from an inherited potentiality of the central nervous system. But this is the point from which studies of development should take off. We should certainly must be content of label such patterns as 'instinctive' and leave it at that. Genes may control behavioural development, but to do so they must interact with the developing animals' environment.

Gottlieb's ducklings need some auditory stimulation if they are to be primed to respond to the maternal calls. It will be our task to discover such factors in the behavioural environment of animals which affect the development of their instinctive patterns. Clearly we must avoid any simple dichotomy which ascribes instinct to the genes and learning to the environment. Both must be involved in the development of all behaviour. This point might not seem to need such emphasis, but in fact misunderstandings here have been, until quite recently, the basis for a considerable dispute among animal behaviour workers.

Development of Behaviour

May studies of the development of behaviour are designed to investigate the mechanisms that underlie behaviour patterns observed in adult animals. Such a study may be relatively direct—for example, determining what factor affect nest-sit selection by a bird species—or it may be a complex investigation—for example, establishing the ontogenetic determinants of feeding strategies in omnivorous rodents. Some studies apply conditions of either deprivation or enrichment; others alter the quality of specific types of stimulation provided during ontogeny rather than vary the overall quantity of stimulation. There types of investigations provide useful information on the development of traits within natural ecological and social context.

Other studies of behaviour development utilize a variety of manipulations to test the effects of various treatments. Examples of these approaches are presented when we examine the chronology of development. First, a bit of elaboration is needed on some aspects of experimental design and testing procedures used by investigators studying the development of animal behaviour.

Aspects of Experimental Design

King set forth seven parameters relevant to the study of early experience. His scheme, or portions of it, can be rather broadly used in designing many of the experiments in behaviour development—particularly those that attempt to assess the effects of early experience on behaviour.

Parameters of the early experience treatment:

1. Age of the animal at the time the early experience treatment is given.
2. Type or quality of the early experience treatment.
3. Duration or quality of the early experience treatment.

Parameters of the tests administered to determine the effects of early experience treatment:

4. Age of the animal at the time of testing
5. Type of test used to assess the effects of early experience treatment.
6. Testing for the persistence of the early experience treatment.

The parameter of the genetics of the species being investigated.

7. Testing different strains or species to determine the effects of early experience on different animals.

Testing Procedures

We can test the long-term influence or persistence of early experience treatments by two types of experimental design. One method provides an animal with early experience treatment and tests for the effects of that experience at various ages; we use the same animal for each test. This type is called *longitudinal design* and requires the use of appropriate repeated-measures statistical tests to analyse the results. Alternatively, we can provide an animal with an early experience treatment and test it only once at a later age, because in a second test of the same animal we would not be able to tell whether the observed effect was due to the treatment or to the experience of the first test.

We therefore avoid re-using the test subject, but possibly we lose some information about the persistence of early experience treatment on that subject. This second type of design is called *cross-sectional*. In some studies of early experience, investigators subject animals to stimulus deprivation or enrichment by manipulating either the quality or quantity of one of three types of stimulation or some combination : (1) sensory, (2) motor, or (3) social.

In every case investigations compare the test performance of animals reared under deprived or enriched conditions with the test performance of control subjects reared under *normal conditions*. *Normal rearing* in this context usually refers to the common laboratory environment—a cage that restricts the amount and type of activity of the animal, contrived social contacts, a prepared diet, and constant ambient conditions. These conditions are extremely simplified when compared with natural environment. Thus, we should keep in mind that the control conditions are often stimulus-poor and not at all comparable with the subject's natural surroundings. In the chronological sequence which follows we provide some general remarks and several examples of investigations of behaviour development at each stage or period.

Feeding and Food Preferences

Is it possible that the food ingested by a lactating female rat influences the later food preferences of her pups ? To test this question Galef and *Henderson* experimented with thirty-two pups born to four female rats. After the birth of the pups, the investigators removed the lactating mothers from their home cages and placed them in separate compartments during three one-hour feeding periods each day. They gave two of the females food prepared by Purina and two food made by Turtox. When untested rats are given a choice test they normally prefer the Turtox diet; the foods differ in taste, texture and colour. *Galef and Henderson* measured the food preferences of the pups for seven days, beginning at seventeen days of age. Figure shows that the young rats reared by lactating mothers fed the Purina food ate proportionately more Purina chow than Turtox; and pups reared by lactating mothers fed Turtox preferred Turtox chow. A rat pup may acquire information about the food its mother is consuming in three possible ways.

Young rats may consume faeces dropped by the mother; they may ingest or smell particles of food adhering to the mother's fur or oral region; or the favour or odour of the mother's food may be transmitted in her milk. Further experiments conducted by *Galef and Henderson* have confirmed that information about diet is transmitted through the mother's milk. The result of these experiments have some interesting implications for our understanding of the development of food habits in young rodents as they leave the natal home site and fend for themselves.

Prior to this work the assumption had been that young rats had to learn to locate solid foods without any assistance from adult rats. However, the study outlined here and other work by *Galef* and his colleagues suggest that young rodents may in fact obtain some of this information via their mothers. At the age of weaning the feeding site selection of young rats is influenced by interactions with adults, and in particular by olfactory cues associated with conspecifics. When presented with a choice, rat pups select a feeding site associated with either conspecifics or their excreta in preference to a clean site. Interestingly, if pups are reared without contact with conspecifics, they select a feeding site without respect to whether it is clean or has conspecific cues present. When pups are reared away from conspecifics, but then are given five days exposure to conspecifics prior to testing, that exposure is sufficient for pups to show a strong preference for feeding sites associated with conspecifics stimuli.

Food Preferences in Turtle

Burghardt and Hess investigated the feeding preferences of snapping turtles (*Chelydra serpentia*). They fed separate groups of newly hatched turtles either horsemeat, fish, or worms for a twelve-day period. When they tested the turtles on day 13, with a choice of three diets, the turtles demonstrated significant preferences for the food on which they had been reared. The investigation then provided each group of turtles with a different diet for twelve additional days and again tested their diet preference among the three food types. The turtles again exhibited clear preference for the food they had eaten during the twelve-day period immediately after hatching. We can draw two important conclusions from these results : (1) the initial feeding experience may be critical in affecting later dietary preferences in snapping turtles, and (2) effects of the initial feeding experience may be irreversible.

Further testing would be needed to confirm and extent these conclusions to the retention into adulthood of the effects of the initial feeding experience, but some type of food imprinting may be occurring—that is, the turtle's food preference may have become fixed even through the turtle with still consume other foods.

Feeding Behaviour of Gull Chicks

The feeding behaviour of laughing gull (*Larus atricilla*) has been studied in both field and laboratory settings. *Tinbergen* and *Perdeck* made observations in the field of the effects of holding cardboard models in front of young gull chicks in their nests. They found that hungry chicks use their bills in a pecking and stroking pattern directed at the bill of a parent, a behaviour that induces the adult bird to regurgitate food for the chick. *Hailman* examined how this food-begging behaviour in gull chicks developed. His results indicate that the behaviour resulted from an interaction of genes and environment.

Table 4.1. Food preferences of snapping turtles after one meal of one of two foods

		First Meal		*Second Meal*		*Choice Test*
Group	*N*	*Food*	*Total Pieces Eaten*	*Total Pieces Food*	*Eaten*	*Number Preferring First-Fed Food*
1	12	Horsemeat	26	Worm meat	19	12
2	13	Worm meat	28	Horsemeat	34	8
Totals	25		54		53	20

The genome of the young gull provides the information necessary for the correct maturation of the bird's sensory and motor systems. As it matures the bird learns to peck at the parent's red bill, which contrasts with its black head. By using cardboard models of a gull's head to test pecking behaviour, *Hailman* found that several days were required to attain a record of 75 to 90 per cent hits. Thus, full development of the begging behaviour requires the genetically programmed development of sensory-neural systems to receive and interpret the environmental stimuli, motor capacities for responding, and the experience associated with learning to peck for the food.

Influence of Environment

Some forms of behaviour do not appear until a particular stage of development is reached. Some of these behaviours seem to develop without any obvious practice. For example, pigeons start to flap their wings and fly erratically at a particular age, and their flying ability appears to improve with practice. However, *Grohmann*, in a classical experiment, reared a group of pigeons in tubes so that they could not move their wings. Another group of the same age was allowed to develop without restraint. When the unrestrained pigeons had reached the stage at which they could fly satisfactorily, those that had been restrained were freed. *Grohmann* discovered that they also were able to fly immediately upon being released. Chickens that are featherless as a result of mutation develop normal wingflapping vestibular reflexes, even though there are completely ineffective.

At first sight it would appear that the behaviour is independent of environmental factors, but we must not assume that the ability to fly develops irrespective of any ontogenetic eventuality. Rather, flying ability is contingent upon some essential conditions, although our knowledge of these id sketchy. A similar example is the vocalisations of pigeons and domestic fowl, which are highly stereotyped and appear at certain stages of development. They are not dependent upon auditory experience but upon certain hormonal conditions that are a normal part of overall maturation. Many movement patterns appear to develop in the absence of practice or example. However, study of the developmental history of some behaviour patterns has suggested that they involve fragmentary and incomplete movements that might influence the course of development.

As an example, we can consider *Kuo's* (1932) work on the development of behaviour in chick embryos. *Kuo* placed windows in the eggshell and was able to directly observe the behaviour of the embryo. He found that the embryo is constantly subject to stimulation both from its own activity and from outside the egg. Various movements can be observed, including initial passive movements of the head caused by the beating heart. Active head movements appear within a few days, and these may be accompanied by opening and closing of the beak. By the seventeenth day, movements similar to pecking are apparent. Some scientists believe that the chick learns to peck in this way. Others ridicule this idea. Whether or not true learning occurs, there id little doubt that the developing embryo responds to stimuli from both inside and outside the egg and that these stimuli may influence development. Thus, *Margaret Vince* showed that in some species, stimuli produced by the embryo can influence the development of other eggs in the clutch and can lead to synchronisation of hatching.

Gilbert Gottlieb shows that duck embryos can respond to maternal calls five days, before hatching and that such experience may influence the duckling's subsequent behaviour. However, other workers provide evidence that movement in the chick embryo develops in the absence of sensory stimulation and that the movement patterns that occur later in ontogeny are not necessarily influenced by earlier movements. Readers should remember that the young of different species develop in very different ways. While ducklings and goat kids are mobile as soon as they are hatched or born, blackbird nestlings and kittens are helpless at this stage. We are not surprised to discover that the postnatal development of the latter is influenced by experience. Should we be surprised at prenatal influences in the case of animals that are still inside the egg or womb at the same overall stage of development? There seems little difference in principle between the protection the duckling receives while in its egg and that the young blackbird enjoys in the nest.

Similarly, there would seem to be little difference in principle between the effects of experience in the two cases. For each species, certain types of experience are essential for normal development for example, adult cat vision is abnormal if there has been any disruption in the co-ordination of the vision from its two eyes during the first few months of life—by inducing an artificial squint or by covering each eye on alternate days so that the two eyes never work together. The exact nature of the experience needed for normal development varies considerable from one species to another.

Learning during Development

Some animals appear to be pre-programmed to learn about certain aspects of the environment during particular periods of their development, for example, language learning in humans. In contrast to other types of learning, language learning is *pre-programmed* in the

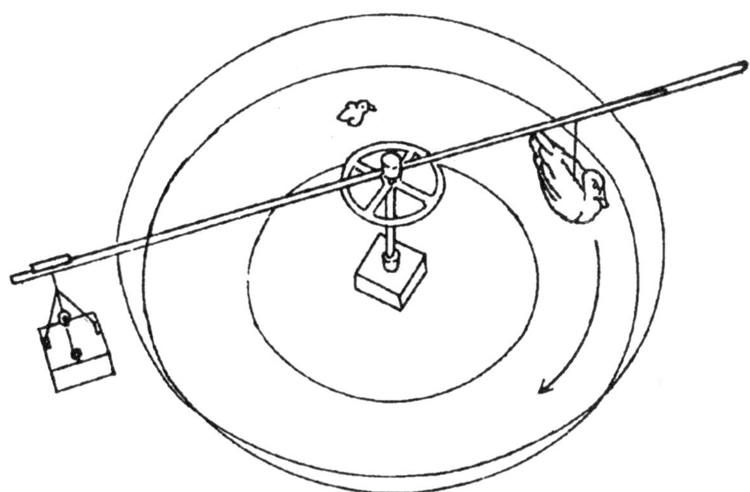

Fig. 4.4. A duckling following its 'mother' in an apparatus designed to test aspects of imprinting.

sense that in the absence of any obvious reward or punishment. For example, the young of many precocial species show a fairly indiscriminate attachment to moving objects. Thus, newly hatched mallard ducklings, separated from their mother, will follow a crude model duck, a person, or even a simple box moved slowly away from them. Some stimuli are more effective that others in eliciting this following response.

In the natural environment the most effective stimuli normally are provided by the mother, and approaches to the mother often are rewarded by body contact and warmth or by food that the mother uncovers. In the laboratory the attachment to a model can be enhanced by food rewards. The more an animal develops an attachment to one object, the less interested it is in others. This process of learning, through which attachment to the mother normally develops, is called *imprinting*. It occurs during a particular, sensitive period of development, which varies according to the species and the circumstances.

Imprinting may have long-term effects, beyond the attachment to a parent or foster parent. In many mammals such early experience affects subsequent social adjustment. In a number of bird species, imprinting has been shown to affect subsequent sexual behaviour. In short, imprinting is a process of learning that occurs at a particular stage of development and that affects subsequent behaviour toward parents, peers, or sexual partners. If the object of attachment is a cardboard box, then the duckling will become attached to the box as to a parent. If male zebra finches (*Taeniopygia guttata*) are raised by Bengalese finches (*Lonchura striate*), then they court Bengalese finch females when adult. Although this type off learning may be influenced by rewards, it is not dependent upon them or upon any particular consequences of the behaviour. The learning is pre-programmed to take place as part of the normal process of development and in whatever circumstances pertain at the time.

Juvenile Events

In Fishes

While considerable fewer studies of the ontogeny of fish behaviour have been conducted than for some other vertebrate groups, there has been some recent progress. Perhaps the most extensive investigations were those of *Ward* and his colleagues. After observations of the orange chromide (*Etroplus maculatus*), a theoretical model for the ontogeny of numerous behaviour patterns was proposed. The arrows marked "+" indicate facilitation; the solid arrows marked "–" indicate inhibition. The dotted arrows, for diagrammatic simplicity, represent environmental feedback to the organism. The environmental factors implied by the dotted arrows are those indicated feeding into the model from the right. The dotted arrows marked "+" are proposed routes of facilitation between behavioural units.

All subsequent behaviour patterns of this species are developed from two initial movements; glancing and micronipping. Data for a number of other cichlid species and for some salmonid species seem to fit this model, though extensive additional testing is needed.

Birds and Mammals

The period which lasts roughly from fledging (birds) or weaning (mammals) until the animal is fully independent and on the way to maturity can be loosely termed the *juvenile stage*. Development is a continuous process, usually marked by various identifiable events, such as birth or fledging,. As investigators we often divide the continuum into stages or periods for convenience. Because animals of different species—and indeed individuals of the same species—often proceed through development at varying rates, the timing and length of these periods will vary. Several pivotal events usually occur during the juvenile period as we define it here, including dispersal from the natal site, puberty, learning appropriate communications signals, various experiences which influence later behaviour patterns, and, for birds, the first migration toward the equator. Play behaviour, which for some species constitutes an important activity during the juvenile period, will be discussed in more detail later in the chapter.

Social Deprivation in Domestic Animals

One method of assessing the possible significance of a particular experience or stimulation during development involves depriving the organism of one or more of these experiences or stimuli and measuring the effects on subsequent behaviour. Some of these treatments may actually commence prior to the juvenile period, whereas others start during this stage. The effects may be noticeable and measure-able during this stage or later in the life of the organism. Deprivation may be social, sensory, or motor, or some combination of these.

Fuller investigated the effects of experiential deprivation on the social behaviour of dogs (*Canis familiaris*). He and his colleagues placed beagle and terrier puppies in isolation for varying lengths of time. They included periodic contact with a human handler or another dog at prescribed intervals in some of the isolation regimes. They then gave the dogs an open-field test and several learning tasks, and observed the subjects for periods when the dogs could make contact with a towel, a ball, or another puppy. The results were similar to the findings for monkeys reported by *Harlow* and his co-workers the more severe the deprivation, the more pronounce the dog's behaviour deficits and abnormalities.

The dogs reared under the most restricted regimes sometimes failed even to leave the starting area of an open-field test, were generally less active than dogs from other treatments, and lost most frequently in competition with another puppy. Prior to Fuller's work two explanations of the observed behaviour differences between deprivation-reared and normal animals had been proposed. Some investigators ascribe the differences to the deterioration of previously organised neural patterns; this theory is sometimes referred to as the *disuse hypothesis*.

Others postulate that the behaviour deficits resulted from a lack of critical information input necessary for proper development of neural patterns and normal behavioural responses; this concept is often called the *loss-of-information hypothesis*. Fuller proposed a third explanation, which he termed the *stress-of-emergence hypothesis*. Animals that have been in isolation and that are suddenly thrust into a test situation loaded with many novel stimuli may suffer from a stimulus overload—that is, they are faced with a multitude of competing emotional responses.

To test his hypothesis *Fuller* suggested three alternative treatments : (1) providing the isolate-reared dogs several brief, pretest exposures to the testing arena to habituate them to the situation; (2) giving the dogs a tranquilisation drug, like chlorpromazine, before testing to reduce the effect of stimulus bombardment; or (3) giving the dog several brief periods of handling by the experimenter before testing. These methods have been tried and, with varying degrees of success, have produced some reductions in the stress of emergence. Fuller's hypothesis appears to be valid in some experimental deprivation studies, but we must consider the two earlier hypotheses as primary explanations for the behavioural deficits that result from isolation treatments.

In Monkeys

From the 1950s onward *Harlow* and his colleagues have conducted a series of studies on the effects of social deprivation on the behaviour of young, adolescent, and adult rhesus monkeys (*Macaca mulatta*). The investigators usually reared young rhesus monkeys under various conditions for periods of time during the first two years of life. The conditions included: (1) Rearing in total isolation in chambers that remove the infant from all social contacts and most external stimulation. (2) Isolate rearing, but with a cloth or wire surrogate mother. (3) Peer group rearing with other monkeys of similar age. (4) Rearing with the mother only. (5) Rearing with the mother, but with varying periods of separation at specified age intervals. (6) Rearing in small social groups in the laboratory (often used as a control condition).

Both the age of the monkey at the time the treatment is initiated and the duration of the rearing experience are critical parameters of social deprivation. This list of rearing treatments is not intended to be exhaustive; researchers have employed other rearing conditions as well as gradations or combinations of these treatments.

When given a choice, most infant rhesus monkeys preferred a cloth surrogate mother to a wire surrogate monther, even when the nursing bottles were attached to the wire mother.

As we might expect, the severity of the observed effects due to deprivation rearing varies with the degree of social isolation and the duration of the treatment. The first five treatment conditions are listed in approximately the order of degree of deprivation. Behavioural deficits (e.g., failure to perform complete behaviour patterns, performing abnormal behaviour patterns, lack of responsiveness to conspecifics) are produced in varying intensities, depending upon the severity of the isolation. These effects include withdrawal, fewer social interactions,

Fig. 4.5. Young rhesus monkey with surrogate mother.

deficiency in understanding communication, and inability of perform normal sexual behaviour. Among the better-known effects of social deprivation are rocking and swaying, self-clasping and other self-directed actions, huddling behaviour exhibited by monkeys reared in peer groups, and poor maternal behaviour and abusive behaviour toward their infants by surrogate-reared females—hence the name "*motherless mother*".

Studies performed by Harlow and his associates have taught us a great deal about the nature and development of bonds between mothers and their infants and between maternal members of conspecific groupings of primates. We know from their research that several qualities of the rhesus mother, such as contact (even that provided by cloth surrogates), warmth, and some types of movement, are important for normal infant development and development of affection. Young rhesus monkeys spent more time clinging to cloth surrogate mothers than wire mothers, even when the source of milk was associated with the wire surrogates. These investigations have given scientists information that should help provide better rearing conditions for primates in zoos and laboratories.

We should note two additional points in connection with these studies of social deprivation. First, some of the behaviours that *Harlow* and others have recorded as *"abnormal"* occur occasionally in rhesus monkeys reared under "normal" conditions: self-directed aggression has been observed on a number of occasions in nonisolate-reared subjects. Another study reported that in free-ranging rhesus monkeys—or those in their natural habitat, but with some spatial restrictions—only 50 per cent of the young born to primiparous females survive to the age of twelve months. *Harlow* and his co-workers have also noted that "motherless mothers" exhibit much better maternal behaviour with their second infant. Part of the deficiency in maternal care recorded in "*motherless mothers*" may result from their need to learn how to be good mothers and not strictly to the emotional abnormalities associated with isolate rearing.

Second, and quite significantly, *Harlow* and his associated have succeeded in socially rehabilitating isolate-reared monkeys. They accomplished this by exposing six-month-old social isolates to three-month-old normal monkeys, called *therapy monkeys*, for two hours per day, three days per week, for one month. The effects produced by some types of early social isolation are not, as was once thought, irreversible, though even rehabilitated monkeys continue to exhibit some behaviour defects.

Singing in Birds

In the past several decades an increasing amount of attention has been given by investigators of animal behaviour to patterns of birds song—their control, development, and evolutionary significance. What about the development of song in birds? There appear to be a wide diversity of developmental strategies leading to the diversity of songs and calls with which we are all familiar. From comprehensive analyses of a number of bird species there appear to be at least two major strategies for song development; (1) imitation of the songs of others, particularly of adult conspecifics; and (2) invention or improvisation. Underlying both strategies are the issues of what type of templates exist, which provide some genetic basis for and help to shape or guide the song learning process, and the possible existence of a sensitive phase for development of the song repertoire. Several examples should help explain song development.

In Marsh Wern

The marsh wren (*Cistothorus plaustris*) has been used extensively in studies of song learning. Male marsh wrens in nature sing over one hundred types of songs; neighbouring males generally sing identical song types; and males often interact by countersinging with one another using the same song type. To test development of song learning males were reared in special housing conditions where the songs they heard could be completely controlled. Males were played specially prepared tutor tapes of songs; each tape contained nine different songs. Males were exposed to one set of tapes for the period 15 to 65 days of age, a second tape for age 65 to 115 days, and a third tape the following spring. The males learned, by imitation, the nine songs on the tapes to which they were exposed prior to 65 days of age, but did not learn the songs on the subsequent two tapes. The males never sang any invented songs. Further investigation revealed a more refined estimate for the peak sensitive period, which falls between about days 35 to 55 of age. In addition, males learned song types played for either three days or nine days. In these additional tests some males improvised some songs, presumably from elements of the songs on the tutor tapes. The learning situation used for these marsh wrens involved only tutor tapes played over loudspeakers.

In another test young wrens were first exposed to a number of song typc vis tutor tapcs and then were given a period of social interaction with adult males with varied song repertoires. The period of social interaction occurred in the fall of the first year for some birds and not until the following spring for others. Data on song repertories for these birds indicate that song learning can occur early from the tapes, or at either of the period of exposure to adult males. Hence, there is some flexibility in terms of song learning. These findings also make it clear that we should be careful in using only rather artificial stimuli like loudspeakers. Social interaction has also been shown to be a critical factor in language acquisition in human children.

In Male Swamp Sparrow

A longitudinal study of song development has been carried out for male swamp sparrows (*Melospiza georgiana*) that were hatched in the wild and brought into the laboratory. Beginning at 16 to 26 days of age the birds were given song training twice per day for forty days with songs typical for the species. Recordings were made of the songs of all birds once each week, commencing shortly after training, when birds were just over three months old, and continued until they were over one year of age. When these recordings were analysed, a seven-stage sequence of song development was discernible. The syllables used in training are shown at the top of the figure. Young birds began by singing what *Marler* and *Peters* call subsong at an average age of 272 days. They progressed through sub-plastic song and began to sing the plastic song at an average of 299 days age. Crystallised song began, on average, at 334 days of age. During the course of this developmental sequence, the duration of the song decreased. Syllabic structures began to emerge during sub-plastic song. Analysis of the syllables revealed that about 30 per cent were imitations of the training songs and 70 per cent were inventions or improvisations.

By the time of emergence of the crystallised song the number of syllables sung was only 23 per cent of the potential repertoire; both imitated and improvised syllables were included in the crystallised song of most birds. During development it was not unusual to record songs characteristic of more than one stage of the sequence on the same day. However, once the singing of crystallised song began, the birds rarely reverted to an earlier stage. *Marler* and *Peters* (1982) hypothesize that the pattern of song development noted for the swamp sparrow may be characteristic of many song bird species.

In Female Mice

Puberty is a critical event in the lives of most organisms—for many, sexual maturation marks the onset of reproductive behaviour and the production of progeny. Puberty is also important in terms of the population biology of many species. For house mice (*Mus musculus*) the deme structure generally involves one to several adult males, three to seven females, and their pups and juveniles. Some juvenile females may disperse from the natal site, whereas other may remain within the deme, but virtually all juvenile males disperse. Within this social context the timing of puberty in female mice has been shown to be significantly affected by various social and environmental factors.

Puberty is measured by the occurrence of first estrus; the heightened levels of estrogen result in cornification of the cells that line the vagina. The phenomenon can be detected by microscopic examination of a vaginal smear. The presence of a mature male mouse or daily exposure to urine from mature males accelerates the onset of puberty in young female mice, as does daily exposure to urine from lactating females and urine from single-caged females, regardless of age, results in delays in the onset of puberty. Mice exposed daily to clean bedding or to urine from single caged diestrus females reach puberty at ages that are intermediate between the acceleration and delay effects produced by exposure to urine from the other sources.

Table 4.2. Mean ages and ranges for sexual maturation of female house mice (Mus musculus) given various treatments with urinary chemosignals

Treatment	*Mean Age at Puberty (days)*	*Age Range (days)*
Control females—exposed to water treatment daily	35	29-42
Females caged with an adult male	27	24-31
Females exposed daily to urine from adult males	31	27-36
Females exposed daily to urine from grouped females	41	36-45
Females exposed daily to urine from estrous females	30	26-36
Females exposed daily to urine from di-estrous females	36	30-41
Females exposed daily to urine from lactating females	30	27-35
Females exposed daily to urine from pregnant females	31	26-36

These effects have been demonstrated in both laboratory and wild stocks of house mice and several of the effects have been replicated in stocks of wild *Mus musculus* maintained in the cloverleaf islands of superhighways. When mice are exposed simultaneously to urine from two or three sources the outcome depends on which sources were used. For example, if treatment involves any exposure to urine from grouped females, puberty is delayed in young test females—regardless of what other sources are used. If all donor sources involve urine which accelerate puberty, then the combination treatments result in earlier maturation, but not any earlier than using one of the sources alone. Diet and daylight have also been shown to affect the timing of puberty in female house mice.

Hence, a variety of contextual cues from the developing mouse's environment can influence internal physiological events. These effects, in turn, have important consequences for the onset of reproductive behaviour in the mice. The chemosignal effects will also influence the length of the interval between a mouse's birth and when it starts to reproduce—called *generation time*. Generation time is a key element in determining the rate of growth or decline in the numbers of young mice entering the population over given length of time.

In Insects

Fruit Flies

Fruit flies (*Drosophila*) have been used in a wide variety of investigations of the development of behaviour; studies have been done both on larvae and on adults utilising field and laboratory conditions. The primary activity of larvae is feeding. As the larva moves across the food surface it provides with its mouthparts, ingesting food with each cycle of extension and retraction of its body. The larva of *D. malanogaster* as adult flies. The rate of feeding reaches a peak early in the third instar and then declines during that instar. Before pupation the larva may migrate to a pupation site away from the food source, or it may remain at the food source during pupation, the pattern varies for different species and sometimes even within a species.

Some species may burrow into the soil to pupate, possibly to decrease the risk of predation. The adults of many *Drosophila* species, particularly females, have been studied more extensively than larvae—adults perform a considerably greater number of behaviour patterns. Two major behaviours related to reproduction undergo developmental changes in females, sexual receptivity and oviposition. *Manning* studied sexual receptivity in *D. melanogaster*. As the corpora allata, which release juvenile hormone (JH), and ovaries grow larger, the female becomes receptive. This generally occurs about 48 hours after the *eclosion* of the adult fly from the pupa.

Immature females reject courting males that attempt copulation. One hypothesis relating the physiological and behavioural events postulates that JH may affect the brain directly or indirectly, lowering the threshold for sexual receptivity. If activity functioning corpora allata are implanted in females at the pupal stage, the emergent flies exhibit earlier sexual receptivity and have larger ovaries than nonimplanted controls. After mating, females become unreceptive again and soon the oviposition behaviours increase. The turning off of receptivity appears due to at least three factors: the presence of sperm in the female' receptacles, the act of copulation itself, and a secretion from the paragonial gland of males. The increase in rates of oviposition is probable due to the increased size of the ovaries, which provide information to the nervous system via stretch receptors, and to the presence of male paragonial gland secretion. *Ringo* discusses the development and maturation processes in females and males of a group of *Drosophila* species inhabiting the *Hawaiian islands*. Development takes a long time in these species, lasting for days or weeks. For males of *D. grimshawi* a series of eight behaviour patterns were observed and recorded at four ages during a one-month period following eclosion.

In general, the diversity of behaviours observed increased with age, and the relative frequency of each behaviour increased with age. Males of this species from leks, in which groups of males display communally and mate with females attracted to the lek. For days 15 and 22 there were pronounced increases in sexual and agonistic behaviour—courting, jousting, and abdomen drag. The increases in behaviour correspond to the time when the males are most likely to be competing for copulations.

Honey bees

Many bee species live in large colonies in which there are castes—physiologically, behaviourally, and often morphologically different forms occurring together. For example, in the honey bee (*Apis mellifera*) and other related species of social bees, there are at least three casts; queens, drones, and workers. Queens mat, lay eggs, are larger, often do not forage or defend the colony, and eat proteinaceous food. In contrast, workers generally do not mate or lay eggs, are smaller, and actively engage in foraging, defence of the colony and nest-building. The eggs of the queen(s) hatch into larvae, and after these pupate they emerge as adults. One key question concerning development is: How and when does the determination of cast occur?

Several factors appear to be important in this process. Which factors are important varies between major groups of bees, but in general they include size of the brood cell, amount of food mass provided to the developing larvae, quality of the food supplied to the larvae, and possibly some chemical cues transmitted with the food. For most species studied, these effects bring immediately upon hatching or very early in larval life. Conditions both within the hive (e.g., loss of a queen) and outside the hive (e.g., changing seasons) can influence the numbers of workers and queens produced. Most worker bees of there social species undergo changes in behaviour as they develop which correspond to changes in their functional roles within the colony or hive. Resting and patrolling occur throughout the twenty-four days surveyed. Many other activities occur in a sequential pattern, beginning with high levels of cell clearing in the early days. This is followed by activities related to the comb and tending the brood. For the last portion of their lifespan the workers became foragers; the average worker bee living in a temperate zone climate survives to the age of six weeks. For many of the activities there is some overlap—bees shift among behaviours off all types on a given day, up to the start of foraging activity. Several physiological changes are correlated with the shifting patterns of functional roles. Young bees have enlarged hypopharyngeal glands. These produce a major component of the bee milk, part of the diet fed to larvae. Levels of secretion for invertase, an enzyme involved in the conversion of nectar to honey, are highest during the middle portion of the lifespan.

Wax glands are small in the youngest workers, but this is followed by a gradual increase to maximum size by about days 16 to 18, and then the wax glands decline rapidly. These changes correspond with the higher levels of comb-related activities during this age range. Many other traits follow similar patterns of correlation between developmental events throughout the life of the bee and its changing functional activities in the hive. Most social bees actively guard and defend their colonies. How do they develop recognition of nestmates and discriminate them from other conspecifics? Work by *Breed* (1983) indicates that bees will use environmental odour sources, or, if the environmental cues are controlled they will use genetically based cues to discriminate nestmates from non-nestmates. Investigators have shown that the cues necessary for making these discriminations are acquired prior to emergence as adults.

Altricial Young of Selected Species of Non-primate Mammals

Altricial young provide a good opportunity to study early stages of behavioural development among mammals that might be more difficult to analyse in the seemingly more complex precocial young that process advanced sensory and motor capacities from birth. Among altricial young, there are relatively distinct periods of thermotactile, olfactory and visual control of behavioural responses and underlying motivational processes, and the relatively simple forms of motor activity can be studied as they become organised into more complex action patterns While, it is true that limited motor abilities may hide advanced sensory and integrative processes in altricial young, it is equally true that advanced sensory and motor abilities may mask simple behavioural capacities in percocial young. The problem is, therefore, to determine, through analytical experiment procedures and theoretical synthesis of the findings, the natural of behaviour organisation at each stage of development. We are not prepared to develop detailed models of the behavioural organisation of young at the various stages but we can sketch the major lines alone which development proceeds. Among altricial young these lines are clear from many good descriptions of the development of behaviour toward the mother, siblings and nest or home site in a variety of mammalian species. Suckling and nom-suckling contact with the mother, and huddling with siblings in the nest, take an early lead in behavioural development among altricial young.

At first the mother takes the initiative in suckling, but gradually the young contribute increasingly to the initiation and termination of feeding sessions, until the relative contribution of each shifts and the young play the dominant role in the feeding interaction until weaning intervenes. Similarly, huddling among siblings is a relatively simple behaviour pattern while the young are about more freely, behavioural interactions among siblings become more complex: young become capable of many more responses to one another and respond increasingly to individuals and their characteristics rather than to group characteristics. Initially, the young are either passively confined to the home area or are actively oriented to it but as development proceeds they become able to move more freely in relation to it at a distance. This early form of orientation to the home area wanes after a period and the young's relationship to its socially conditioned environment is based upon its relationship to its companions and to various specific functions that emerge as development proceeds.

Suckling

Among all species of altricial mammals suckling is initiated shortly after birth. The newborn actually begin to nuzzle the mother's nipple region during parturition, when given the opportunity in the intervals between births. Kittens, however, are only rarely successful in attaching to nipples and this is very likely the case in other altricial species. *Ewer's* suggestion that the delay of an hour or so *post partum* in the onset of suckling in kittens is based upon the absence of a special releasing stimulus for nipple grasping seems unnecessarily elaborate.

Kittens have little opportunity to áttach to nipples during parturition even when the mother is not actively giving birth. The mouth is engaged in a variety of activities such as licking the kittens, eating the placentae, licking birth fluids from her fur and the delivery site, etc., which make it difficult for kittens to attach to nipples. Not until parturition is completed does the female make herself available for feeding. The position assumed after parturition by the exhausted mother among cats, dogs and other small mammals facilitates nipple searching and grasping by the newborn, her limbs outstretched forming an oval-shaped corral in which the young are confined. Comparable positions are assumed by rat, hamster, and rabbit mothers. Most successful artificial mothers have been modelled after the immediately post-partum nursing female.

In this position the mother presents a variety of attractive thermal, tactile and olfactory stimuli. The effects of these stimuli on early suckling have been investigated through the presentation of one or several of the component stimuli under controlled conditions in artificial brooders. Among the important features of the mother's body that stimulate approach and nuzzling appear to be the fur-textured surface, and the rounded shape with few entrapping crevices or projections that distract newborn from locating the nipples and surrounding areola. Body surface temperature in the range of 32 to 39°C, as measured in various species, and as provided in artificial mothers, has proved attractive to newborn rats, rabbits, puppies and kittens.

The moist surface of a pulsating plastic tube was most attractive to brooder-reared rat pups in the study by Thoman & Arnold. In addition to being attracted to the mother's nipple region by these stimuli; newborn kittens are led along certain paths in their nuzzling of her body by patterns of thermal and tactile stimuli. *N. Freeman* found a broad but well-defined thermal gradient extending over the entire body surface of the lactating cat that could provide a basis for locating the nipples in newborn kittens. When on a floor surface at 30.2°C a thermal gradient extended from the mother's limbs, which had a surface temperature of 33.3°C, to her nipples which were at 37°C. The pattern of fur growth offers additional directional stimulation : when the mother is lying on her side, crawling against the grain leads kittens upward from the midline to the lateral nipples. Moreover, the areola surrounding the nipples, with its bare, warm moist surface, attracts kittens.

It has been suggested that the pattern of hair growth on the belly of the pig provides paths which piglets follow in finding nipples. Static features of the mother and more active ones, such as her licking that steers kittens to her ventrum, offer attractive stimuli which initially bring newborn into contact with her. More detailed and patterned features lead to them to the nipples.

Suckling and Thermotactile Stimuli

Thermal and tactile stimuli provide the basis for the earliest suckling approaches to the mother in kittens, rat pups, and newly born rabbits and puppies. In the kitten, the snout and lips are supplied with low threshold thermotactile receptors and greater anterior-end sensitivity to such stimuli appears to be the general rule among newly born altricial young. The snout functions as a probe as it moved by side-to-side head movements during crawling, a characteristic of kittens and other altricial young that move in this manner. These movements enable the newborn effectively to scan a fan-shaped region in front of it.

Contact with a low-intensity tactile or thermal stimulus slows the return head swing and the newborn veers in the direction of the stimulus by repeated movements of the contralateral forelimb and cessation of ipsilateral forelimb movements. On the other hand, contact with a strong tactile stimulus (e.g., wire texture floor surface, or excessively warm or cool stimulus (e.g., overheated or cooled floor surface or overly warm or cool littermate, initiates head withdrawal followed by cessation of forward movement and veering away.

If the stimulus persists, the newborn pivots a half circle and crawls away from it. This withdrawal response, which can also be elicited by a puff of air directed at the newborn's face, has been used to establish early escape and avoidance learning in newborn of several altricial species. At this early stage, thermal and tactile stimuli in the litter situation also stimulate newborns to activity which eventuates in their crawling to the mother. *Okon* has shown the distress-inducing effects of strong tactile stimulation and extreme thermal conditions upon hamster, rat, and mouse pups, as measured by a high rate of audible and ultrasonic calling. This builds up during the first weeks after birth then declines at the end of the second and beginning of the third week. Although rough handling or tail pinching are most effective in eliciting ultrasonic calling, it is particularly significant that loss of contact with tactile stimulation (i.e., even nesting material) stimulates calling in three days old rat pups and young hamsters.

Similarly, cooling caused by exposing three to nine day old pups to a 22°C ambient temperature stimulates a significant rise in the proportion of animals that vocalise and in the rate of ultrasonic vocalisation. Newborn rabbits are stimulated to activity at 20-25°C ambient temperatures which cause a decline in body temperature of 9°C.

Among puppies and kittens loss of contact with tactile stimulation from the mother and littermates stimulates an increase in activity that usually consists of pivoting and calling. In puppies, quiescence is restored when head and body contact is made with a soft tactile stimulus, and a most effective stimulus for quieting kittens that have lost contact with the mother and litter is the weak tactile stimulation of the face and forehead provided by a canopy of soft bunting. Cooling is as effective as tactile stimulation in a arousing puppies and kittens to activity and in stimulating calling.

In an ambient temperature of 30°C puppies remain quiescent, breathing lightly in a relaxed sleeping posture, but if the *ambient* temperature falls to 27°C or rises to 32.5°C, they become alert and restless, being to call, and exhibit rapid breathing. Kittens placed on a cool surface (18°C) become active and call during the first week. Many crawl to a warmer area if this is available, but others become so active they simply walk off the table surface. Newborns that lose contact with the mother therefore become active as a result of both the loss of tactile stimulation and exposure to a lowered ambient temperature. They begin to pivot and circle, emitting calls to which the mother often responds by bringing her body closer and into contact with the young or by licking them.

The newborn is likely to make contact with the floor surface around the mother which is warmed by her body and with her extended fore- and hind-limbs as well as the warm furry surface of her nipple region. These ensure that the newborn will regain contact and warmth and if it has not fed for a time, that it will initiate nipple-locating movements and finally suckle from the mother.

Nipple Grasping and Sucking : Specialised Responses

Nipple grasping and sucking are specialised responses of the newborn to features of the nipples and the surrounding areola. The specific nipple characteristics which enable newborns to grasp them have not yet been established for any altricial species and so newborns do not initially attach spontaneously to the nipples of artificial brooders. The projection of the nipple from the surrounding mound of bare skin (areola) is an important feature enabling kittens to locate and grasp the nipples. While nuzzling the mother's fur they crawl upward on her body but upon reaching the areola region their nuzzling abruptly changes to gentle nose tapping and forward crawling stops. As they contact the projecting nipple with the nose and lips, the head is withdrawn and raised and the mouth is opened; the nipple is grasped by a forward head lunge with mouth open, and often several such lunges are made before the nipple is centered in the kitten's mouth. This is a variable response even in newborn kittens.

Brooder reared kittens develop a different pattern in adapting to the longer and more flexible rubber nipple used. They approach in a similar way but upon making contact they position the nipple at one side of the mouth, move the head sideways, thus bending the nipple, then allow it to spring back into the mouth. If brooder nipples are placed in rounded depressions on the brooder surface, kittens find it difficult to attach to them. The reason appears to be that when probing the depression, the kittens are stimulated more strongly around the face by the edges of the depression than on the nose and mouth by the nipple. Face stimulation stimulates crawling and strong pushing into the depression but appears to inhibit a response to the nipple.

The location of the nipple on a mound, as in the mother, therefore has the opposite effects: crawling is inhibited by the nose stimulation and nipple grasping is elicited. Each kitten rapidly develops a preference for either a single nipple or a pair of nipples. Texture differences on the surface surrounding the nipples may play a role in this discrimination., since R.J. Woll has shown, in brooder-reared kittens, that a discrimination between differently textured nipples flanges, one providing milk and the other without milk, can be learned by the second or third day. In rat pups tactile stimulation of the upper lip appears to be the stimulus for nipple grasping. Pups, whose upper lips had been desensitised, failed to grasp the nipples although they nuzzled them when placed on the nipple region of anaesthetised mothers. When they were placed on nipples, however, they sucked normally.

Once the nipple is grasped and sucking brings, young are extremely sensitive to tactile and perhaps thermal stimuli within the oral cavity. A series of postural changes during suckling has been described in rat pups, in which changing nipple stimulation and finally milk ejection stimulate first treading then stretching responses. Treading anticipates actual milk ejection and may be in response to subtle oral tactile or thermal stimuli resulting from milk let-down within the mammary gland.

In human mothers, oxytocin released in response to infant crying raises breast temperature by 1°C as a result of milk let-down. The likelihood of thermal stimulation playing a role in nipple grasping arises from *N. Freeman's* finding in the lactating cat that nipple temperature is highly (37°C) than the nearby surrounding skin (34.7°C) and G.H. Rose's observation that warming the nipple and flange of an artificial mother aids the newly born kitten to find and grasp the nipple. Stanley and his associates have shown that the newborn puppy is able to modify its sucking in accordance with the rate and pattern of milk flow, probably on the basis of oral tactile stimulation.

Suckling and Intraorganic Stimulation: Early Stage of Development

The initial and terminal of suckling cannot be explained entirely by reference to exteroceptive thermal and tactile stimuli from the mother even in the early stage of development. Young of all altricial species exhibit a periodicity of feeding that appears to depend on internal sources of stimulation generally labelled, in aggregate, as hunger motivation. The most obvious during suckling is the milk which fills the newborn's stomach. It has been assumed that filling of the stomach terminates suckling as a result of stimuli arising from muscular distention of sensory receptors in the stomach, and, conversely, that an empty stomach stimulates the young to initiate suckling and to be responsive to nursing stimuli. *James* was the first to show that in young puppies (i.e., less than 26 days) suckling was not inhibited by preloading their stomachs with adequate amounts of milk. Latencies to initiate suckling and the duration of suckling were not different in preloaded and non-preloaded puppies which had been separated from their mothers for up to 3 hours, although preloaded puppies consumed less milk.

Further study showed that preloading to the extent of overloading (i.e., milk regurgitated through the mouth) could inhibit suckling and reduce milk intake, but strong effects had to be made with these fully preloaded puppies to keep them awake. Even though they attached to nipples, once they were released after being hand held and stimulated they released the nipples and fell asleep. Thus, the stomach probably does not directly regulate the responsiveness of young puppies to the exteroceptive nursing stimuli of the mother. *Hall* has recently shown this even more clearly with rat pups during the first 10 days, using the anaesthetized mother to test for sucking.

Hall deprived one group of pups of suckling for 22 hours and removed a second group from the mother just before testing and anaesthetisation. Both deprived and non-deprived pups were placed on the mother's exposed nipple region and their readiness to attach and suckle was measured. There level of general activity was also measured. During the first 10 days, deprived and non-deprived pups attached with the same latencies. Latencies for attaching decline in both groups from just below 200 seconds to between 50 and 75 seconds; activity levels did not differ substantially between the two groups for most of the 10 days period.

This study shows that readiness to suckle was not dependent upon any differences in stomach loading that resulted from differences in the recency of suckling. Anaesthetised rat mothers do not release milk in response to sucking because of the inhibition of oxytocin release necessary for milk let-down. In Hall's study, therefore, pups sucked from dry nipples yet they continued to suck for the remainder of the five minutes test period. In this early period, therefore, ingestion of milk is not an essential stimulus for sucking; Hall reports that pups eight days of age and younger suck for at least six hours, continuously, despite the fact that they obtain no milk throughout this period. Similar findings have been described for kittens during the first three weeks after birth. Two groups of kittens were studied : one was presented with a normal lactating mother and the other with an anaesthetised non-lactating female in whom sucking did not, of course, cause milk release.

Both groups were fed adequate amounts of milk by tube directly into the stomach at four hour intervals and were exposed to the 'mothers' for six hours each day, the exposure starting two hours before a scheduled tube-feeding. During the first three weeks the dry-sucking group sucked as frequently and for as long as the group that obtained milk during sucking. Both groups sucked (i.e., were attached to nipples) for between 31 and 42 minutes of the first hour, indicating that latencies to attach were rather short and did not differ between groups.

At the end of two hours, both groups were tube fed according to schedule and the effect was to reduce sucking during the third hour, equally in both groups, but not eliminate it. This study suggests that milk intake is not essential either for the initiation of suckling during the first three weeks or for maintaining sucking. We shall see that in puppies, rat pups, and kittens, the situation changes as the young become older and milk intake plays an important role in sucking. Yet there is a periodicity in feeding even during the early period that is not based entirely upon the mother's behaviour since it appears in brooder-reared kittens as well. Among kittens and rat pups the young sleep during sucking though they may remain attached to the nipples, and when the mother raises to leave, some are so strongly attached that they are dragged from the nest. The ingestion of milk plays a role in sucking but its role is indirect. Following stomach filling with milk the young may be lulled to sleep, as a result of a sharp reduction in general arousal perhaps through inhibitory effects of stimuli from the distended stomach.

Sucking may be terminated, therefore, or fail to be initiated or only weakly initiated, in young preloaded with excessive amounts of milk. During the first 10 to 15 minutes after preloading with milk, kittens exhibited an overall lethargy that made them unresponsive to all forms of stimulation and preclude any response to a nursing mother. There was then gradual recovery and the kittens became more responsive to stimuli and often began to initiate suckling approaches with nipple grasping and sucking. Suckling has many intraorganic effects some of which derive from milk ingestion and others from exteroceptive stimuli received during contact with the mother.

Hofer and his associates have uncovered an effect of milk intake on cardiac function. Two week old rat pups, separated from their mother overnight, show a decline in heart rate that cannot be attributed to a decline in body temperature. These investigators have shown that only tube feeding at regular intervals (i.e., one or four hour intervals) prevents this decline : *Koch & Arnold* have extended these findings to pups four to 14 days of age. In addition they have found in four to 10 day old pups that even non-suckling contact with the mother by tube-fed pups accelerates cardiac functioning more that tube feeding alone. Since rat pups in this age range are unable to thermoregulate, the effect may be based upon the thermal insulation provided by contact with the non-suckling female as well as contact with a source of warmth well above the pups' body temperature. Rabbit young suffer a loss in body temperature when isolated in 20-25°C ambient temperature and contact with the mother produces an immediate rise in the young's body temperature. This suggests that suckling exerts an important influence on the body temperature of young partly through body contact and partly through the warming effects of milk: full body contact with the mother has similar or even greater effects. It is perhaps significant that others from a different perspective have proposed that maternal behaviour among mammals has evolve primarily in relation to the thermoregulatory needs of altricial young and that nursing is a secondary development which evolved later.

Nevertheless, a number of investigators have shown that altricial newborn of a number of species can learn discrimination of various sorts on the basis of milk reinforcement of sucking. Tactile, taste, and olfactory discriminations have been established and sucking has been 'shaped' by the presentation of milk. If the effects of reinforcement are related to the underlying motivational conditions, then in these young, reinforcement must act in a different way than in adult animals with well-defined motivational systems. The nature of this difference, however, is not yet known.

Suckling and Olfaction

Olfaction enters early into the suckling pattern of altricial newborns but it may be several days before it plays a significant role in the approach to the mother and in nipple searching. As early as three days of age, rat pups show evidence of responding to the odour of a lactating female or of nest deposits, and on the fifty day pups show behavioural arrest and EEG synchronisation in response to material odour. By the ninth day the evidence of response to maternal odour is clear. Bilateral bulbectomy on the second day has a severe and irreversible effect on suckling in rat pups, resulting in a gradual weight loss beginning one or two days after surgery and 80% mortality by the sixth day. Pup approaches to the mother when she enters the nest and interactions with littermates are only mildly affected by bulbectomy. At later ages also, bulbectomy produces a severe reduction in suckling and consequent weight loss, with high mortality but mild effects on social interactions with the mother and the littermates. The effect of bulbectomy appears to be specific to nipple grasping since approaches to the mother and nuzzling in her fur are not different in bulbectomized and normal or sham-operated control pups. This suggests an early role for olfaction in nipple grasping. *Kovach* and *Kling* similarly found that bulbectomized kittens could not suckle with the mother but were able to suckle when placed on an artificial nipple. The rabbit pup, kitten and puppy show evidence of use of olfaction in their responses to the mother during the first week. Artificial odours have been applied to the mother's nipple region and later tested alone to determine whether, in the course of suckling, young rabbit pups and puppies are influenced in their behaviour by the presence of the odour.

Rabbit young thus exposed to odours during the first five days of suckling began to respond positively to the odour alone on the second or third day and puppies exposed similarly responded immediately when tested for the first time on the sixth day. Among kittens any of the three functional pairs of nipples may be suckled by each of the kittens of a litter during the first day *post partum*. By the second and third day, however, each kitten confines its suckling to either a single nipple or a pair of nipples. Once nipple position preference are established nose contact with non-preferred nipples does not elicit nipple grasping as it did on the first day.

To determine whether olfactory cues could be the basis for these preferences, brooder-reared kittens were presented with two nipples and flanges, one of which provided milk while the other was blind.

Two-day old kittens were tested for their ability to establish as olfactory discrimination. Within a day or two the kittens were able to discriminate between two artificial odours: they approached and suckled almost exclusively the nipple-flange combination with the positive odour. The ability of rat pups to approach the mother on the basis of an olfactory stimulus has been clearly established by *Moltz and Leon*. From 14 days pups respond to the odour of lactating mothers in preference to that of non-lactating females.

The odour emanates from a volatile component of the mother's faces which the young ingest, and if mothers are fed different diets the offspring can distinguish the odour of their own mother. Under these conditions, the odour of a strange mother has no special attraction for them, including that material odour acquires its attractiveness through experience. Olfactory stimuli appear gradually to become a component of the young's approach and suckling response to the mother in these altricial species. From the beginning, olfactory stimuli are able to influence the newborn's behaviour. Ammonia fumes (which may in fact have irritating tactile effects) stimulate pivoting and vigorous crawling in newborn rats. There are, undoubtedly, odours to which newborn respond positively from birth onward, particularly those related to birth fluids etc, but, in the main, odour appears to acquire their significance through association with thermotactile stimuli that accompany them. When the development of olfactory responses to the mother's body and particularly to those regions involved in nursing, certain changes occur in the suckling pattern.

1. Suckling becomes more specifically a response to the newborn's own mother and to specific aspects of the mother's nipple region (e.g., development of nipple position preference).
2. Since olfactory stimuli from the mother may reach the young before actual contact is made with her, the young may initiate the suckling approach while the mother is still at a short distance and thus they may anticipate her approach.
3. Since olfactory stimuli may be spread outside the nest or home site, familiarity with maternal odour may enable young to widen the range of their activity to these areas.
4. Since olfactory stimuli may also be deposited on littermates, young may also begin to respond to their siblings on this basis.

Suckling and Vision

At the time vision becomes functional in altricial young, the suckling pattern is well established on the basis of olfactory and thermotactile stimuli. These stimuli, however, can only be received by young when the mother is close by or in actual contact with them. Use of vision enables young to perceive the mother at a distance and approach her for suckling. The onset of visually stimulated suckling behaviour is usually taken as the beginning of distant approaches to the mother.

In the rat this begins between the fourteenth and sixteenth days, in the hamster around the second week, in the puppy around the fifteenth day, in the rabbit around the tenth to fourteenth days, and in the kitten around the seventeenth day. It must be kept in mind, however, that vision may begin to function earlier in suckling but at short distances where it would be difficult without specific examination to distinguish between a thermotactile, olfactory, or visual basis for the young's approach. While there is considerable descriptive evidence of visually guided approaches to the mother beginning sometime after eye-opening in the species we have been discussing, experimental evidence if often lacking.

In the kitten eye-opening occurs around the seventh to ninth day and there is evidence of the onset of visual functioning between the fourteenth and seventeenth days. Visually guided approaches to the mother from a distance start around the seventeenth day and are fully established by the twenty-first. In the 10 day old rabbit pup an odour associated with the mother initiates approach to the mother which is then guided by visual stimulation. The implication which is then guided by visual stimulation. The implication of the studies on maternal pheromonal stimulation in the rat by *Molta and Leon* is that, from around the fourteenth day, pups identify a lactating female by odour but their approach to her is guided by visual stimulation.

Altman and his colleagues have shown that rats orient to the mother and littermates in the home cage, presumably on the basis of visual supplement by olfactory stimuli, during the third week. The beginning of visual functioning in the rât pup has been found on the fourteenth day by *Turner* who studied pups' ability to find an exit door, marked by a vertical stripe, in a circular field.

Pups improved rapidly from the fourteenth day onward in their ability to go from the centre of the field directly toward the door, a distance of 13 inches (33 cm). Visually guided approaches to the mother in puppies described by *Rhein-gold* as starting around the fifteenth day and increasing therefore, are correlated with visually evoked responses in the cortex which assume the adult form at around the same age. By four weeks of age, puppies will approach a two-dimensional drawing of a female, paying particular attention to the head and flank, and at a slightly older age to the mammary region, when they are hungry. The onset of visually stimulated approaches to the mother marks an important stage in the young's development with implications for the organisation of suckling behaviour.

It is no doubt significant that the ages at which young increasingly initiate suckling approaches to the mother from a distance in the rat pup, puppy, and kitten correspond roughly with the ages at which intraorganic stimuli becomes increasingly important in the regulation of feeding. If the young approaches the mother when she is unresponsive or not easily available it is likely to result in either failure to suckle or in rejection by the mother. It is during this stage that maternal rejection or evasion of the young become prominent. Sucking approaches made when the mother is receptive result, on the other hand, in her co-operating with them by assuming a nursing position. Young gradually become sensitive to subtle features of the mother's behaviour indicating her readiness or unwillingness to nurse and adjust their suckling approaches accordingly. Moreover, each young approaches the mother to suckle individually, whereas earlier the mother's approach had stimulated the entire litter simultaneously and all young suckled together. Once nursing is initiated with one offspring other young often join in, but each young acts on the basis of its own perception of the situation rather than to the stimuli that initiated suckling in its sibling.

Suckling and Intraorganic Stimulation: Late Stage

Between the tenth and fourteenth day an important change takes place in the intraorganic stimulation of suckling among rat pups. Around this age the latency to initiate suckling (with an anaesthetised mother) begins to differ between pups that have been deprived of suckling for 22 hours and those that have only recently been removed from their mothers. Latencies are very short for the deprived pups but range above 200 seconds for recently fed pups. Moreover, failure to receive milk from the anaesthetised mother, which had little effect on maintenance of suckling earlier, now begins to affect suckling: pups shift from one nipple to another, within the five minutes suckling test, presumably in response to the dry nipple. Similarly, around the third week of age, kittens suckling from a dry mother initiate nuzzling of the mother's nipples with the same frequency as kittens suckling from a lactating mother, but the frequency of attaching and suckling declines rapidly from the earlier period and the duration of sucking also declines to 20% of the previous period. Meanwhile kittens suckling from a lactating mother continues to suckle for extended periods.

After 30 days of age, preloading the stomach of puppies with a milk-chow mixture has the same effect of reducing or eliminating further eating from a dish as the same amount of food consumed voluntarily by control puppies. This contrasts with the earlier finding that preloading the stomach with either milk alone or a similar mixture has no effect on suckling. Indications are, therefore, that arousal produced by intraorganic stimulation has become more specific than earlier so that exteroceptive stimulation received before or during suckling is not sufficient to initiate or maintain suckling. Milk intake and its intraorganic effects have become an essential part of the sucking pattern.

At this stage of development, stomach filling or its absence has become a more focal source of arousal than earlier and it determines to a large extent the young's response to the exteroceptive stimuli associated with suckling. The kind of exteroceptive stimuli that aroused suckling approaches undergo a change during this period. Among rat pups this is thc pcriod during which suckling approaches are increasingly initiated by the pups from a distance, in response to olfactory and visual stimuli from the mother. Among kittens there is the beginning of kitten-initiated suckling approaches to the mother, starting around three weeks of age.

At this stage suckling approaches are increasingly determined by the past experience which the young has had with the mother and less by immediate stimuli for suckling. This was shown in a study by *Kovach and Kling* in which they found the kittens reared in isolation from an early age until about three weeks and fed by stomach tube to prevent suckling, were unable to initiate suckling when returned to their mothers. While they attributed this to the waning of the suckling reflex, our analysis indicates that at three weeks off age the sucking reflex is only a small part of the suckling pattern.

Moreover, as part of an organised suckling pattern it initiated mainly on the basis of intraorganic stimuli rather than the previously effective proximal exteroceptive stimuli. In a study comparable to that of *Kovach and Kling* kittens were reared in isolation from one to three weeks of age with feeding by suckling from an artificial rubber nipple mounted on a brooder. When these kittens, in whom suckling had maintained as an active component of the feeding pattern, were returned to their mothers there was a delay of about 20 hours in initiating sucking. The previously isolated kittens had no difficulty in establishing sustained contact with the mother, which in all cases took less than three hours and in most kittens was accomplished within the first hour. Their difficulty lay in perceiving the mother as an object to be suckled. They crawled over her body, slept in close contact with her nipple region and often nuzzled, but not until they had been without milk for many hours did they finally exhibit nipple grasping and suckling.

In these brooder-reared kittens the suckling pattern had been organised in relation to the brooder and artificial nipple, which differed in many ways from the mother. Under comparable conditions in the brooder situation, these kittens would initiate suckling within seconds after being returned to it. As young develop, therefore, the relative influence of exteroceptive and intraorganic factors in the regulation of suckling shifts in favour of intraorganic factors. Moreover arousal undergoes changes: It can no longer be described simply in terms of general arousal. Arousal has become more specific as a result of experience in the litter situation and of neutral maturation which gave rise to new physiological relationships between visceral and somatic sensorimotor processes. When aroused, young are channelled into particular patterns of behaviour not only by the prevailing exteroceptive stimulation, as earlier, but by intraorganic factors among which previous suckling experience plays a crucial role.

Huddling and Related Behaviour

Newborn huddle from birth onward : after nursing has been completed and the mother leaves the nest or home site the young crawl into contact with one another and form a closely knit huddle. Since heat loss is dependent upon body surface exposure to the environment in young with poor thermoregulation, the huddle maintains body temperature and metabolic activity. Huddling is an early form of social behaviour among littermates that may be closely related in development to later forms of group activity (e.g. play).

Huddling and Thermotactile

Cosnier and Alberts have shown that early huddling among rat pups is based upon thermal and tactile stimuli provided by littermates. Pups up to 10 days of age are more attracted to a warm live pup (i.e. 37°C) than a cool dead one (25°C) but the warm pup can be replaced by warm tubing or a warm fur-lined nylon sleeve. Pups are particularly stimulated to huddle when the surrounding floor is cool. *James, Welker and Crighton & Pownall* analysed the behaviour of individual puppies, when separated from their littermates in warm and cool environments, in an attempt to understand the thermotactile basis of huddling. Separation stimulates pivoting and vigorous crawling until the puppy makes head contact with its littermates.

The essential feature of the contact is thermal, however, since contact with cooled puppies or contact with littermates in an overly warm environment elicits withdrawal rather than huddling. Within the huddle temperature is maintained at 30°C and puppies remain quiescent, breathing quietly and in a relaxed sleeping posture. Warming the huddle to 32.5°C causes dispersal as each puppy responds to contact with its neighbour by withdrawing; cooling the huddle intensifies huddling as each puppy responds to the warmth of its neighbour by approaching it. The greater thermotactile sensitivity of the face and snout over the remainder of the body assures that puppies face into the huddle. The few observations of kittens indicate that similar factors operate. The role of tactile stimuli in huddling receives some confirmation from a study by *Scott et al.* (1974) in which puppies that were isolated in a warm environment but nevertheless vocalised were induced to become quite by confirming them in a section of stove pipe the floor of which was lines with soft material. The tactile stimulation, and perhaps the insulation provided, are similar to that provided by neighbouring puppies during huddling.

Huddling and Olfaction

Singh & Tobach (1974) found that bulbectomized rat pups, from the second day onward, tended to be apart from the little for longer periods than their normal littermates, although in general their social behaviour was not particularly deviant. The suggestion that olfaction plays a role in huddling was more clearly established by *Alberts* (1974). Using intranasal infusion of zinc sulphate to make pups anosmic he found that huddling was completely eliminated in pups 10 days age and older, but was retained by five day old pups. *Fox* has shown that olfaction may begin to play a role in the puppy's response to its littermates as early as the first week after birth.

An odour applied to the mother's body, and therefore spread to the littermates, elicited strong approach response in puppies exposed to the odour during the first five days of life, but strong avoidance responses from puppies presented with odour for the first time on the sixth day. Similarly, among rabbits, material odour, which is shared by the littermates, elicits a strong searching response when held close to the young's nose without tactile contact as early as the second to fifth day after birth. Evidence suggests also that hamster pups may begin to respond to litter odours at the end of the first week.

Orientation to Littermates and Vision

Huddles become somewhat less compact as young develop thermoregulation and less in need of insulation from heat loss. Nevertheless, they continue to aggregate in groups which may consist of several young instead of the entire litter as earlier. Of greater interest is the fact that they orient to littermates from a distance suggesting that vision is beginning to plat a role. As late as 17 to 20 days of age, about a week after eye-opening, puppies still do not orient to littermates in the home cage from a distance. This does not mean that they do not react to them but they continue to show circular movements similar to those seen earlier. Starting around 25 days of age, however, they begin to approach littermates from a distance directly.

Rheingold and Eckerman showed that kittens respond to littermates at a distance even when they do not orient directly to them. They studied the earliest age at which kittens would be comfortable by the presence of a littermate in a strange environment. At two weeks of age, the earliest age at which testing was done, kittens emitted 50% fewer vocalisation when a littermate was present compared with when it was absent. Despite the fact that only sketchy evidence is available to trace the development of huddling and other relationships with littermates among these altricial species, the main outlines of this development are clear.

Stages in the development of responses to littermates are similar to those found in the development of suckling : there is an initial stage of dependence upon thermotactile stimulation, a later stage of olfaction based responses and a final stage of visually guided responses to littermates. Moreover, early responses to littermates are dominated by exteroceptive stimuli while later stages show evidence of an increase in self-initiated approaches to siblings based upon the developing social relationships. Huddling is a relatively simple behaviour pattern compared to suckling, but it represents only the earliest form of interaction among littermates that ultimately develops into the complex patterns of play.

Development of Home Orientation

The nest or home site is the socially conditioned environment in which the young undergo the early stages of their development. The young, from an early age, develop a pattern of home orientation that plays an important role in their relationship to the mother, and the organisation of their behaviour in relation to their surroundings and littermates. Since the original discovery of home orientation in the kitten orientation to the site has been found in the hamster, rat, and puppy, in one form or another, suggesting that further research will reveal that this is a widespread and basic phenomenon among altricial newborn.

Thermotactile Basis of Home Orientation

In all altricial species the nest or home site usually provided with a soft insulating material and is warmer than the surrounding regions, largely as a result of warmth from the mother's body, the huddling young, and the insulation. *Tobach et al.,* found that three day old rat pups were more active in shavings, the usual nest material used in laboratories, than on the surrounding bare floor, on which they remained immobilised. The soft material used to line the nest site in many species is likely to be more attractive to newborns than coarse material because of its tactile properties, as the studies on newborn puppies by *Stanley et al.,* have shown. While these tactile properties are attractive they are of little use in orientating to the nest from a distance; they function mainly to keep the newborn in the nest or to bring to rest a newborn that has been wandering. The situation with respect to the thermal stimulation provided by the nest or home site is quite different.

N. Freeman measured the floor temperature at various regions of the home cage of a mother cat and litter just after they were removed in preparation for testing kittens in home orientation. The floor temperature of the home site was 35.3°C at its centre, 33.6°C at its border, about 25 cm away from the centre, and it decreased from 25.2 to 23°C at a distance of 15 cm. This thermal gradient persisted for some time after removal of the litter, but, of course, if the mother and litter remained in the home region it would be maintained constantly. On the basis of the thermal gradient kittens could distinguish the home from the adjacent corner when placed there during a test. They vocalised loudly and moved about actively in the adjacent corner while they soon became quiet and fell asleep when placed in the home site.

Odours are not involved in this response initially, since the thermal gradient produced by a heat lamp on a floor surface that had not been occupied by the mother and litter or any other cats had the same effect. The existence of the thermal gradient between the home site and the adjacent corner also enabled kittens of a slightly older age to crawl to the home from this outlying region. Also, when tested on a thermal gradient ranging from 15.6°C to 48.3°C over a distance of 56 cm, kittens up to the age of one week or so, crawled in the direction of the higher temperature if they were placed at a region below their preferred temperature. Similarly, starting at one day of age and continuing through the eighth or ninth day, hamster pups orient to the warm end of a thermal gradient, crawling distances as great as 45 cm in one minute.

Okon has shown that hamster pups are distressed in an ambient temperature of 28°C and emit audible and ultrasonic vocalisations. They also become responsive to a thermal gradient and crawl rapidly to the region of their preferred temperature of 33°C. The age at which thermal orientation wanes in hamster pups is variable and depends upon ambient temperature. As pups become older (i.e. 11 days and older) cooler temperatures are required to stimulate thermal orientation: at warm ambient temperatures, still below their preferred temperature, thermal orientation may wane as early as the sixth day. It is clear that thermal orientation is based in part upon a contrast between the pup's body temperature and the thermal gradient and also between the floor surface under the pup and the floor temperature of a neighbouring region. In rat pups the ability to orient to a thermal source develops over the first two weeks with an important change occurring around the end of the first week.

Until the age pups react to contact with a thermal source by remaining in contact with it but they do not orient to it from a distance. This is in agreement with Cosnier's findings, based upon tests in which pups were placed directly in contact with the heat source. Between birth and the fifth day, pups become increasingly active when the ambient temperature is reduced. If they make contact with a warm area they come to rest but there is no indication that they orient to the warm area from a distance. *Alberts* reported that wandering in response to cool ambient temperatures was the basis of huddling among five day old pups and *Fowler & Kellogg* have confirmed this.

After the fifth day, pups will orient toward a heart source provided they can detect it from a distance through a thermal gradient. *Fowler & Kellogg* found that during the first five days pups exposed to an ambient temperature of 21-23°C for one hour declined in body temperature to between 28 to 29°C, indicating failure of thermoregulation with respect to the usual body temperature of pups at an older age (i.e. 36-37°C). Although body temperature decline during this early period, thermal orientation was not evident until after the fifth day. To test thermal orientations *Fowler and Kellogg* (1975) placed rat pups between two chambers only one of which was warned. Between the starting point and the warmed chamber, located about 15 cm away, floor temperature increased from 23 to 36-37°C with a 1°C increase as close as 5 cm from the starting point. The unheated chamber was at 22°C, the ambient temperature during testing. Thus, in effect, the pups followed a rather steep thermal gradient (i.e. 14°C over 14 cm) in crawling from the starting point to the warmed chamber.

In a less steep thermal gradient pups are able to turn in the direction of the warmed floor surface but are unable to follow it for any appreciable distance. Thus orientation to the nest or home site arises early in development among altricial newborn : they respond to lowered temperature (within limits) by calling and increased locomotion. At the earliest age they are poorly equipped to respond to thermal gradients but they do respond to entering or being placed directly in a warmed region: they become calmed, vocalisation ceases, they come to rest and usually fall asleep. The activation caused by cooling is reduced by warmth. At a slightly older age they respond to cooling by adopting a path along a thermal gradient, leading to the warmer region. This ability improves over the next period but, as we shall see, it often gives way to orientation based upon olfaction. However, the calming effect of warm temperatures remain an important feature of the nest of home site beyond the period when orientation to these regions is based upon thermal gradients.

Home Orientation and Olfaction

The nest or home site differs from the surrounding regions in the odours deposited there by the mother and litter. These too provide a basis for orientation to the home site. Olfactory orientation to the home site develops after thermotactile orientation has been established and it may be based upon this earlier form of orientation. Towards the end of the first week, a warmed floor surface is not sufficient to quite kittens and a thermal gradient between two cage regions, lacking home cage odours, does not result in kittens adopting a path from the cooler to the warmer region.

Only the presence of home cage odours quiets kittens and only if the home site has the usual odours and, presumably, an olfactory gradient, can kittens find their way back to it. From the seventh day onward, home orientation in kittens is based mainly upon olfaction, although at the home site, the combination of olfactory and thermal stimuli is more effective than either alone. In the hamster thermal orientation declines as orientation based upon odours from the nest becomes established. Starting around the seventh and eight day, hamster pups turn toward nest shavings and away from fresh shavings when placed on a screen above the border between the shavings. They spend 80% to 100% of their time crawling in the nest shavings. Shavings from nests inhabited by a mother and litter are preferred to shavings from nests inhabited by either a male or a non-pregnant female, and nest shavings from four day old litters are as effective as shavings from older litters. Earlier, hamster pups can distinguish between odours, preferring, for example, the odour of cedar to pine shavings; this appear on the third day.

Nest odours must require an additional period of four or five days to become associated with thermotactile stimuli from the mother and littermate to provide a basis for orientation to the nest. As indicated earlier, rabbit pups and puppies begin to respond to odours from the mother before the end of the first week and it is likely that the same odours deposited in the nest would provide a basis for home orientation. It is clear that the odour acquires its attractiveness through association with other stimuli from the mother during the first five days. Home orientation in puppies that appears to be based upon olfaction has recently been reported by *Scott et al.* Puppies were placed either in their own nest boxes, emptied of the mother and littermates, or in a similarly constructed nest box that was lined with fresh towelling; vocalisation were recorded in each.

One precaution was taken, which, however, may have affected the results: to exclude the possibility that puppies might respond to the thermal difference between the litter nest box and the unused one, both boxes were warmed to about 29°C before testing. Despite this, vocalisations were more frequently in the unused nest box than in the litter nest box from around the sixth to eight day in several litters and on the eleventh-twelfth day in all litters. Vocalisations were infrequent in both conditions at first but they rose sharply after the twelfth day in the unused nest box. The distress caused in puppies by placing them in a strange nest box is similar to the distress caused in kittens by placing them in a strange cage: the rate and intensity of vocalisations rises during tests, and with age they become more frequent and more intense. The locomotory behaviour of the puppies in the above study was not described but, in an earlier paper with older puppies, vocalisation was accompanied by an increase in activity in a strange pen and presumably the same occurred here.

Rat pups exhibit the first signs of orientation to the home sage from a distance of less than 30 cm on the third day and even more clearly on the fifth and sixth days. Placed on a 11.4 cm diameter circular platform, located midway between the home cage and a strange cage, pups point themselves toward the home during most of a three minute test but they are not yet able to crawl to it. Between the ninth and twelfth days pups become better able to locomote over long distances by crawling.

On the ninth day, pups placed in a neighbouring cage reach the home cage in 50% of the tests by passing through a narrow alley, with latencies ranging from two to three minutes. Latencies to traverse the 15 to 30 cm distance between the starting chamber and the home cage became shorter on the twelfth day (i.e., one to two minutes) and 85% of the pups are successful and on the sixteenth day, around the time of eye-opening, nearly all pups orient to the home with latencies ranging from 30 to 40 seconds. At 19 and 21 days latencies are 10 seconds or shorter. Neither the paths taken to the home cage nor the sensory cues utilised by pups were analysed in the above studies.

Turkewitz reported that successful orientation from a neighbouring cage to the nest site in the home cage by 9 to 12 day old rat pups was accomplished indirectly through wall-hugging. The long latencies to reach the home cage in the above studies suggests that a similar mode of home orientation occurred at 12 days of age. Old pups, however, appear to adopt direct paths to the home cage and as a consequence latencies are considerable shorter. On the basis of *Turkewitz's* study, it would appear that olfactory and perhaps thermal stimuli are involved in home orientation by rat pups before their eyes open on the fifteenth to seventeenth day. The long latencies suggest pups adopt an indirect path to the home and this is more characteristic of young that orient on a non-visual basis: the shorter latencies after eye-opening certainly contrast with the longer latencies earlier. Findings reported earlier, that rat pups show evidence of olfactory discrimination of nest shavings as against fresh shavings around the ninth day and respond to maternal odours by the fourteenth day, would also support this interpretation of the above home orientation.

Home Orientation and Vision

The onset of vision marks the beginning of the decline of home orientation among altricial young. With the ability to view the home from a distance it is no longer necessary to return to it: instead young tend to remain in the vicinity of the home or nest site for a period before they disperse.

Moreover, orientation to the home declines also because the young begin to follow the mother and !ittermates and these became centres of orientation rather that the nest or home site. There is, however, a short period after eye-opening and the beginning of vision when home orientation continues and vision is used to enable the young to return to the home from a distance. In the rat, as we have seen, the latency for reaching the home from a nearby region is reduced to a few seconds shortly after eye-opening. This indicates that the young begin to take direct routes to the home, and this probably depends on visual stimulation.

The hamster pup shows a decline in olfaction based orientation beginning around the thirteenth day, shortly after eye-opening, and it is completely absent by the nineteenth day. This may be based upon the gradual increase in visually based orientation. Among kittens the use of vision in home orientation begins around the fourteenth day. Testing kittens in the dark results in some decrement in performance with indications that the kittens are seeking the visual stimuli to which they are accustomed. However, they are still capable of using olfaction and most of them reach home. If olfactory cues are removed, some kittens are still able to reach the home and this increases after the seventeenth day. The use of vision in home orientation among kittens eventually leads to the decline of this behaviour at around the eighteenth to twenty-first day. Instead of returning to the home region, kittens look forward it, or they may actually enter the home region briefly then leave it.

The ability to see the home often enables kittens to wander more freely around the entire cage since the home region is always in view. However, when the home was made especially prominent visually by placing a fur piece there, home orientation persisted until 45 days of age. With home odours absent, kittens began to show this visually based home orientation around 25 days of age gradually perfected it, traversing the age, a distance of about 76 cm, in less than 20 seconds as they moved directly toward the home and came to rest entering it.

Behaviour Development of Altricial Young

The early behaviour development of altricial young can be divided into three stages: in the following section we shall analyse these stages and the transitions from one stage to the next.

First Stage: Thermotactile Stimulation

The neonates' earliest behaviour is organised predominantly in relation to low intensity thermotactile stimulation provided by the mother, littermates and nest or home site. These elicit from the newborn either general or specialised approach responses. The general response of forward crawling is elicited by centrally applied head and face stimulation over a broad area and turning is elicited by laterally applied stimulation, through the close relationship that exists between anterior end stimulation and movement of the forelimbs in an alternating paddling motion and the hind-limbs in a joint pushing action. Specialised responses are elicited by localisation stimulation of the snout, mouth, and lips: the specialised responses take their character from the peripherally organised pattern of movement characteristic of the local region and the restricted locus of stimulation. The thermal and tactile characteristics of the mother, littermates and nest or home site provided patterned thermotactile stimuli which elicit from newborn the general and special approach responses necessary for the initiation of suckling, and huddling, and for remaining within the confines of the home. These responses are only loosely patterned sequences at first, depending upon kinds of stimulation the newborn encounters in succession as it crawls forward.

Stimulated initially to approach the mother by contact with the attractive thermotactile stimulation of her body, the newborn subsequently encounters the more detailed stimuli of the mother's fur around the nipple, the areola, and finally the nipple itself. At this early of development the central control of behaviour is not yet strongly developed beyond the mediation of the admittedly complex sensorimotor integrations underlying general and specific approach and withdrawal responses to stimulation. It is doubtful whether, for at least a short time after birth, the functionally distinct behavioural responses to the mother, littermates and nest or home site are represented by equally distinct central regulatory process. The underlying similarities between the behavioural adaptations to the mother, littermates, and nest or home site, arise from the fact that they share in common responses to similar thermotactile stimuli.

The observation differences arise from the fact that in each of these responses the neonate also responds to specific stimulus properties of the mother, littermates, and the nest or home site. While the relative contributions of central regulatory process and peripherally elicited responses favour the latter in the neonate (as compared with older animals in which central regulation plays the dominant role) central processes are not without effect on the neonate's behaviour. Exteroceptive stimuli (i.e., thermotactile) appear to have two concurrent effects upon the neonate. They arouse the neonate to activity, as when the mother's licking activities the sleeping neonate to raise its head and begin to crawl. Once the newborn is aroused, stimuli also elicit approach responses of a general and specialised nature.

Thermotactile stimuli are therefore both arousal inducting and response-eliciting. Neonates that have been aroused are more sensitive to exteroceptive stimuli that follow and exhibit more vigorous responses to these stimuli. The central component of exteroceptive stimulation therefore is an arousal that in turn potentiates the responsiveness of peripheral processes. Early in development central arousal is relatively non-specific in its afferent and efferent relationship and this contrasts with the specificity of the peripheral responses that are elicited by thermotactile stimuli. It is this contrast between the greater specificity of peripheral responses—their direct relationship to thermotactile stimuli—and the non-specific character of arousal and its contribution to early behaviour, that has led many to characterise the neonate's behaviour with some justification as simply reflexive in nature. Early suckling among newborn rat pups and puppies exemplifies the relationship between central and peripheral processes in the regulation of feeding.

The compelling effect of thermotactile nipple stimuli in eliciting sucking is shown by the fact that prior feeding or stomach loading cannot prevent the young from grasping the nipple and sucking. In the rat pup, if milk is not forthcoming sucking may continue indefinitely; suckling does not continue indefinitely in the kitten but its duration is not shorter that of kittens that obtain milk. The difference may rectify the fact that the rat pups tested with anaesthetised mothers while the kittens were tested with awake, non-lactating females. Yet normally rat pups and kittens do not suckle interminably once they attach to nipples. After a period they began to loosen their grip on the nipples and slide off, or they are detached from them when the mother rises to leave.

The intake of milk has an effect on the peripherally organised response of sucking, but the effect appears to be mediated by a lowered arousal, to the point of sleep. All overt activity ceases, not only suckling, and newborn's behaviour alternates between sleep and feeding for a considerable period after birth. Preloading of newborn with milk has an effect on suckling when the amount preloaded is sufficient to induce sleep. With effort, even under these conditions, suckling can be elicited by bringing the newborn's mouth to the nipple but the central effect of intraorganic stimulation arising from the distended stomach or sensory receptors in the stomach, usually prevails and the newborn releases the nipple soon after grasping it. Preloading also reduces milk intake but the basis for this is not clear. Varying levels of central arousal appear to influence the responsiveness of newborns to exteroceptive stimulation: low-level arousal, that characteristic of newborn shortly after feeding dampens the effect of exteroceptive stimulation, while very high levels of arousal, following long periods without food or upon exposure to low ambient temperature, produces an excessive response to exteroceptive stimulation.

There is therefore an optimal level of central arousal at which the newborn exhibits its typical responses to exteroceptive stimuli. The source of central arousal in the newborn are interoceptive as well as exteroceptive, as studies on stomach preloading indicate. These sources have not been studied to any great extent, but interoceptive stimuli have a history in prenatal ontogeny that predates exteroceptive stimuli. Exteroceptive stimuli may in fact exert their influence, in part, through their effect upon interoceptive processes, as for example in responses to ambient thermal stimulation. During the earliest postnatal stage there is a close relationship between those exteroceptive stimuli (i.e., thermotactile) to which newborn are most responsive in their behaviour toward the mother, littermates, and nest or home site, and the conditions which stimulate it to activity. Thus, newborns are stimulated by loss of contact with the mother or littermates and by displacement from the nest or home site; in the latter cases it is probably exposure to lower ambient temperatures than are normally present in these situations which stimulates, the newborn as studies on kittens have show.

At this early age many different objects having attractive thermotactile properties may be equivalent in their arousal and calming effect upon newborn because the special properties of the mother, littermates and nest or home site do not yet play a role. This fact has allowed investigators to ignore the social nature of the newborn's responses to species mates and to the socially conditioned home and to introduce special criteria for when the young's responses are to be considered social. Since the usual criteria are based upon visual responses to species mates, responses that appear during the third stage of early development are more likely to be called social responses and social behaviour is therefore viewed as a later development. Such a view ignores the ontogenetic basis of social responses and their relationship at each stage to the behavioural capacities and behavioural organisation of the young. The earliest changes in the newborn's behaviour are the formation of extended action patterns, incorporating the earlier hesitating and variable crawling approach response and the specialised feeding and huddling responses into more smoothly co-ordinated patterns.

In brooder-reared kittens, after a day or two of feeding, the kittens crawls to the correctly textured flange (i.e., that association with a nipple that gives milk) in a smoother more patterned movement, with short pauses at crucial points such as the edge between the brooder and the floor and between the brooder and cover and the textured flange. At there points the kitten sniffs the brooder cover and rubs its snout against its surface as it does when it reaches the flange. A short period of flange contact is followed by nipple localisation and nipple grasping and sucking. Among puppies approached to the nipple can be associated with either soft-or hard-textured paths with gradual improvement in turning toward one or the other and heading rapidly for the nipple, grasping it and sucking.

Repeated experience with either warm or cool thermal stimuli along alleys leading to a nipple results in a gradually more rapid crawling approach to either. The formation of these specialised action patterns based upon thermotactile stimuli indicate progress along several lines: approach responses become specialised in relation to the young's discrimination among thermotactile stimuli. In this respect approach responses are formed more rapidly when a low intensity, approach-eliciting stimulus is the positive one than when a stimulus that is initially either weakly approach- eliciting, or is actually withdrawal-eliciting, is used as the positive stimulus. In addition, components of the approach response are integrated into a pattern that appears to be less dependent upon continuous guidance by exteroceptive stimulation and more dependent upon central regulation.

In suckling approaches, striving toward the nipple indicated an early appearance of anticipatory responses in advance of actual contact with the nipple region. These changes channel the earlier general arousal along specific lines. Thermotactile stimuli are no longer equally effective in eliciting approach responses: the newborn turns away from certain thermal stimuli and becomes highly excited when it makes contact with others. There are indicators in the study of Bacon (1974) that negative stimuli do not acquire inhibitory effects at this stage but rather that positive stimuli acquire heightened arousal effects. These may have their origin in the heightened arousal that occurs when the young are stimulated by milk sucking. At this stage, new sources of disturbance may arise when the usual incentive is absent or is altered; this is in fact evidence that action patterns of a broader nature already exist in the young and that they are highly specific to the stimulus conditions in which they were formed.

Second Stage: Olfactory Stimulation

Although in altricial young the second stage of early behaviour development is characterised by the growing influence of olfactory stimulation, its main feature is the increasing specificity of behavioural responses in relation to the familiar objects in the environment. The young use olfactory stimuli to differentiate between familiar and unfamiliar object. The initial exposure to olfactory stimuli occurs during responses to thermotactile stimuli, but in the beginning olfactory stimuli can provide little guidance to the newborn. Except when it encounters strong aversive olfactory stimuli, which induce withdrawal responses, olfactory stimuli cannot elicit either general approach or specialised responses in the newborn. Not until responses have begun to become organised into specific action patterns based upon thermotactile discrimination, and the central arousal processes are channelled, can olfaction begin to play a role. The suggestion is that olfaction advances further the process of discrimination among objects, along lines that are necessarily specific to the mother, littermates and nest or home site.

More important, however, is the ability of olfactory stimuli to arouse the initiation of action patterns that are then guided by combined olfactory and thermotactile stimuli. Olfactory stimuli may therefore determine whether or not a response will occur. There is associated with the onset of olfaction the appearance of olfactory orientation behaviour by means of which young explore their olfactory environment. *Welker* has described the development of sniffing as an olfactory exploration pattern in the young rat, and *Komisaruk* has added important details to the analysis of olfactory exploratory behaviour in this species. When placed in a new environment young initiate non-specific olfactory exploratory behaviour which soon gives way either to specific behaviour patterns upon identification of the odour or to further tactile exploration with the vibrissa. These exploratory patterns range from air sniffing to sniffing of the floor surface and approach and sniffing of objects in great detail. Thus, for example, by sniffing the floor surface kittens find their way to the home region and rat pups by sniffing the air are able to locate the mother at a distance of more than 30 cm. Depending upon the distribution of odours, therefore, olfaction enables young to develop distance perception of objects and to initiate movements towards objects before actual contact is made with their thermotactile properties. In this sense, therefore, olfaction plays a large role in developing central control of action patterns begun with respect to thermotactile stimuli.

Anticipation of forthcoming stimulation and the associated action pattern, exhibited during the latter phase of the first stage of development, is gradually transformed into self-initiated approaches to objects with anticipatory action patterns almost entirely centrally aroused, ready to be performed when the object is reached. Rat pups deprived of olfaction during this phase appear much less oriented to significant social stimuli and to the nest site (*Singh & Tobach*, 1974). The stage is set during this second stage of early behavioural development for the incorporation of vision into the central control of action patterns during the third stage of early development. As olfaction begins to contribute to central arousal processes and the action patterns to which they give rise, it begins to play an increasing role as a motivating condition and as a possible source of distress.

Rat and hamster pups, kittens and puppies begin to show distress when removed from their familiar olfactory environment. In kittens and puppies this is evident at the end of the first week and it appears at a slightly older age in rat and hamster pups; it provides the basis for these young to initiate movement toward littermates, resulting in huddling, and toward the home or nest site during the development of home orientation. Moreover, olfaction serves as a goal for these behaviours in the sense that upon reaching the familiar olfactory situation in the huddle or in the home region, young are calmed and soon come to rest. It is apparent, therefore, that not only perceptual and motor processes become more complex as development proceeds but motivational process share in this growing complexity.

Third Stage: Visual Stimulation

Vision greatly enlarges the young's capacity to differentiate perceptually between the significant objects in its environment. Neither thermotactile nor olfactory stimuli are able to convey to specific location of their nest or home site. While olfaction may indicate that the mother is nearby it does not indicate what she is doing, while vision conveys both. This stage is marked by an acceleration in the young's development of social interactions based upon vision. Initially social interactions consist of visual approaches to species mates at a distance, at which point action patterns based upon olfactory and thermotactile stimuli take over : vision therefore contributes little to the character of the interactions bur does contribute to their occurrence. Thus, for example, kittens at three weeks of age, approach the mother at a distance for suckling but when they reach her, they adopt their earlier mode of nipple searching with their eyes closed.

Gradually, however, vision comes to trigger not only the approach to the mother, but also those action patterns that were formerly triggered by non-visual sensory systems. Kittens at this stage, approach the mother and reach up from beneath her as she remains standing, locating and grasping a nipple without any significant preliminary nuzzling. Brooder-reared kittens walk directly to the nipple and grasp it instead of following a path of nuzzling on the brooder surface, and cage-reared kittens walk directly across the cage from the diagonal corner to the home region instead of crawling along a path that passes through the adjacent corner. Vision accelerates the process of increasing central regulation of behaviour and with this furthers self-initiated behaviour with well-defined goals.

Perceptual differentiation and the multiplication of action patterns, and their growing complexity, imply higher specific central stages of arousal and complex interactions, both facilitating and inhibitory, among these different stages. During the period of transition from olfactory to visual control of behaviour in the young, interoceptive stimulation appears also to become more influential and specific in its effect upon the central state of arousal. At this age, *Hall* found that food-deprived and non-food-deprived rat pups began to differ in their suckling behaviour, the former being faster to initiate suckling with an anaesthetised mother and abandoning suckling if no milk was forthcoming.

Similar findings have been reported in three four week old puppies and kittens whose suckling approach to the mother was influenced by stomach preloading with relatively small amounts of milk. The goal-directed character of behaviour at this stage is exemplified by the persistence kittens show in following the mother around the cage, while they look for an opportunity to suckle. At the slightest pause, they immediately reach up to her nipple and if they are detected by her movement they resume following her.

At other times, when tired, they walk across the cage to join another kitten that has settled in a corner to sleep and huddle against it, falling asleep. While they remain highly responsive to environmental stimulation, their behaviour is not directly elicited by this stimulation: rather, they are capable of making perceptual discriminations among the various stimuli and which stimuli they respond to depends upon the central state operative at the time. During this third stage of development kittens and puppies that are placed in novel environments exhibit signs of distress (i.e., vocalisation and agitated movement) that are relieved when littermates or the mother are placed in the same environment.

Rat and hamster pups orient to the mother and littermates when displaced to a neighbouring, strange cage. Thus, the absence of familiar visual stimuli becomes a source of disturbance that motivates young to region visual contact with social companions. Often the experiment serves a similar role if he has become familiar to the young. Since visual stimuli are likely to be a principal source of social stimuli from this age in into adulthood, and social responses of a clearly recognisable nature are particularly evident during this third stage, this stage had been singled out by *Scott* as the beginning of true socialisation. This analysis has shown, however, that social responsiveness arises almost immediately after birth and continues to grow in depth and complexity during each stage of ontogeny. Visually based social responses have their ontogenetic origin in earlier non-visual stages and they retain their close relationship to these earlier stages throughout life.

Functional Aspects

The neonate's dependence upon thermal and tactile stimulation in its behavioural responses to its surroundings is based upon its needs to maintain an optimal thermal environment for adequate physiological functioning in the face of its inability to regulate fully its own body temperature. Contact with the mother'' body causes a rise in body temperature and huddling with the littermates, shown by *Albert* to result in minimal exposure of the group to ambient thermal conditions, maintains body temperature, and reduces metabolic activity. *Leonard* has suggested that overall rate of heat loss is the stimulus to which newborn hamsters respond in the thermotaxic orientation, and the effect of prolonged exposure to low ambient temperature on ultrasonic and audible sound emission by rat and hamster pups neonates respond to altered thermal conditions, particularly during nipple searching, indicates that thermal sensory receptors located in the snout, lips and face also play an important role. Neonates whose behaviour is organised in relation to thermotactile stimulation are necessarily confined to the close proximity of eat sources and their tactile representatives, and they must possess means of reaching these heat sources from short distances if they are displaced.

The behaviour of neonates is therefore in spatial scope and in the range of thermal conditions under which they can function adequately. As these thermal limitations are relaxed with the development of thermo-regulatory mechanisms, neonates become capable of extending the scope of their functioning, both spatially and with respect to environmental conditions. Olfactory stimuli play an important role in this process. They arise from substances deposited by the mother in the vicinity of the home or nest and have special significance for the newborn, since they have been experienced in conjunction with thermotactile stimulation from the mother, and therefore they can provide means for the newborn to extend its scope of activity. The mother creates an olfactory zone around the nest or home (as well as in the home itself when she is absent) in which the newborn can function because of its growing responsiveness to olfactory stimulation.

Moreover, odours are species and individual-species: they provide a basis for the specialisation of responses to particular kinds of animals and particularly to the individual mother and littermates. Evidence of olfaction-based individual attachments by neonates have been reported for rats and kittens and will, very likely, be found in many different altricial species. Home or nest orientation develops in relation to the differential distribution of odours in the vicinity of the nest or home site. It requires a responsiveness to gradients of olfactory stimulation or to orientated deposits which are the products of the mother's own position in relation to the home or nest as the centre of her maternal activity. The maturation of vision increases even further the scope of the young animal's activity and the period of combined olfactory and visual functioning ensures that early visual functioning will be in relation to the most significant features of the young's social environment and within the zone of its socially conditioned physical environment.

The specialisation of the young's responses to qualitative features of stimuli (rather than simply to quantitative features of the earlier phases) advances further with the introduction of vision. A wider range and greater variety of social signals are displayed visually than through odours or thermotactile stimuli. Play arises during the period of visual functioning and is based largely upon the young's responses to the variety of visually perceived actions on the part of siblings and other familiar contemporaries. Earlier sensory systems do not, of course, fall out of use when newer ones mature and become functional. They do, however, play a different role in behaviour than during the phase of their predominance.

Play Behaviour

Play behaviour has been defined and characterised in many different ways. *Fagen* defines play as "an inexact term used to denote certain locomotor, manipulative and social behaviours characteristic of young (and some adult) mammals and birds under certain conditions in certain environments." We can characterise play in various animals and the groups according to the actions of the animals and the contexts in which the play behaviour is observed. Most investigators recognise at least three types of play, with some overlap. The first type is social play as exemplified by wrestling, chasing, and tumbling activities of the young of many species. A second type provides exercise for developing muscles, locomotor patterns, and other movements. The third type is often labelled "diversive exploration."

Generally, this type of play involves sensory inspection of an object followed by extensive repeated manipulation of the object. Play behaviour of one or more of these types has been recorded in mammals, ranging from rodents and bats to bears, cats, elephants, and whales. Play has also been recorded from a large number of avian species, including raptors, passerines, aquatic or oceanic birds and parrots. To date only a few anecdotal bits of evidence exist regarding play behaviour in other vertebrate groups or among invertebrates. However, additional observations must be made before reaching any firm conclusions limiting play to birds and mammals.

Functions

Investigators have ascribed a wide range of functions to play behaviour. In the most general sense, play functions as practice for adult activities. Animals perform many actions in the course of play which contain elements of behaviours seen in later adult life. Perhaps aggressive behaviour is the best example. As young cats or primates engaging in various forms of social play, their mock attacks, chases, and mild, non-injurious bites are all partly practices for *real life* use of these same patterns a few months or years later. A second function often ascribed to play behaviour is to aid in the process of maturation-growth and development. As young foals or lambs cavort about, alone or in small groups, they use their muscles and develop co-ordinated movements. A third function of play, may be to gain information about the environment; this would be particularly true of diversive play. By exploring and manipulating objects found in their environment, young animals accumulate information that may prove useful later in life.

Finally, play can function in the establishment of social relationships with peers and adults. Some of these may be purely affiliations, whereas others could be related to the establishment of dominance-subordinance relationships. For many primate species and some canids, the aggressive play observed in young juveniles gradually become more intense and adult-like. The patterns of dominance established in play encounters as juveniles can be retained in later adult life.

Canids

One animal group for which play behaviour has been studied extensively is the canids *Bekoff* studied social play and play soliciting in coyotes (*Canis latrans*), wolves (*C. lupsus*), and beagles (*C. familiaris*). His observations indicated some species differences and some age-specific trends for several aspects of play behaviour, for example, play soliciting and agonistic behaviour. *Bekoff* notes that the beagles displayed an early onset for play soliciting and exhibited very little agonistic behaviour during play—on fighting occurred at all, only mild threats. Wolves showed moderate levels of play soliciting, with an increase at the last age interval recorded in the sample. Like the beagles, wolves displayed low levels of agonistic behaviour, consisting primarily of threats. In contrast, coyotes exhibited higher levels of agonistic behaviour through the observation period, and a correspondingly lower rate of play-soliciting actions.

Coyotes generally establish dominance relationships through fights at an early age; this may account for the differences between these animals and the other two species. When all play behaviours were summarised, the beagles were seven times more playful than the coyotes and three times more playful than the wolves. It is interesting to speculate on the possible relationship between these differences in observed play behaviour and the differences in social structures of the species tests. Wolves are generally social, group-living animals, though some individuals may lead a solitary existence for some periods.

Beagles are at least somewhat social animals, though we rarely observe such domesticated canids in semi-natural or natural situations

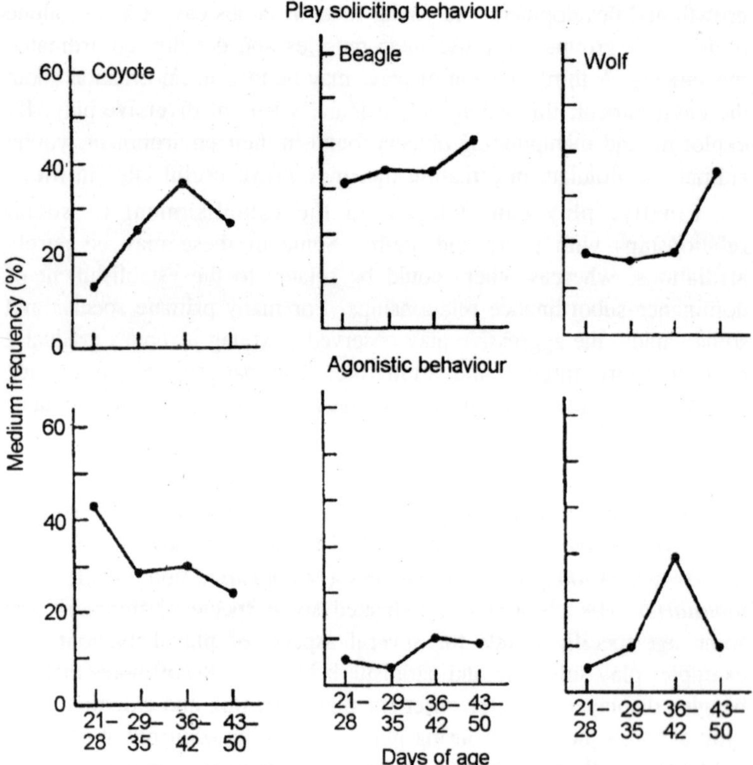

Fig. 4.6. Play-soliciting and agonistic behaviour in canids.

where their feral social structure can be fully recorded. Coyotes, on the other hand, are generally much more solitary in their social organisation, with each animal roaming a large home range. In a recent review *Price* (1984) has suggested neotony as an explanation for the observation that some domesticated breeds of animals are more playful than their wild ancestors. Man has selected for the retention of juvenile characteristics in adulthood. Thus, the beagles in the foregoing examples may be more playful than the wolves because of the effects of man's selection.

Keas

Keas' (*Nestor notabilis*) are large parrots that inhibit parts of New Zealand. A variety of play behaviours have been observed and reported for this species. The most extensive observations were those of *Keller* on captive keas in zoos. The young keas perform a variety of acrobatic maneuvers, including somersaults, sliding on snow-covered slopes, and hanging by their bills from tree branches or hanging upside down by their feet. Social play in groups is also quite common and involves wrestling and activities which can be described anthropomorphically as resembling "*hide-and-go-seek*" and "*king of the hill*". The solicitation to play in a social situation may involve such tactics as adopting a posture normally used in defence with the head down and one foot raised or the stiff-legged walking posture common in some mammals and a few other birds. Last, keas engage in a great deal of object play. They manipulate objects of all sorts using both feet and bills, toss items into the air and fly at them, and they play with snow when it is present.

Human Children

Numerous studies have been conducted by both developmental psychologists and ethologists on play behaviour in young *Homo sapiens*. *Hutt and Bhanvani* conducted a follow-up study of earlier work by *Hutt* in order to attempt to make some predictions about differences in play behaviour based on longitudinal sampling. A young child confronted with a new toy will first investigate and inspect it (specific exploration) and then play with the toy (diversive exploration). Three to five-year-old nursery school children can be classified into three mutually exclusive categories with regard to their responses to a new toy: non-explorers, who visually inspect the toy, but do not handle or play with it; explorers, who thoroughly investigate the toy, but fail to do more than that with it; and, inventive explorers, who investigate the toy and then play with it in a variety of innovative ways. Do these individual differences in exploratory behaviour provide any predictive insights with respect to traits in the children at a later age?

To test this question *Hutt and Bhanvani* obtained data for about fifty children from the original sample of one hundred used to generate the three categories above. The children were seven to ten years of age at the time of the second sampling procedures. Each child was given a series of tests to measure creativity and a personally questionnaire. Each child was rated by parents and teachers on behaviour, adjustment and development. The results provided support for several suggestive conclusions—really more like hypotheses for further testing:

1. Children who has been more creative and imaginative in their early play behaviour were more likely to be divergent in creativity at the later ages; this was particularly true for boys.

2. Lack of exploring was related to later lack of curiosity and adventure in young boys, and to difficulties in personality and social adjustment in young girls.

Hutt and Bhanvani note that some of these effects may be attributed to early childhood differences between the sexes; boys are more exploratory in their play activity for a longer period of their life, and girls are socially and linguistically more advanced than boys at nursery school age. We might also note that further longitudinal follow-up testing to assess longer term effects of these early differences would be both appropriate and necessary to strengthen or extend these conclusions.

5

FIGHTING BEHAVIOUR

Agonistic behaviour can be defined as behaviour that appears to be intended to inflict moxious stimulation or destruction upon another organiser. The notes of intents necessary to exclude such destructive behaviours as a person accidentally stapping on an out. A brief, generally accepted, definition of this class of activities can be obtained from *Hinde* who suggested that "attaching, threat, submissive, and fleeing behaviour from a complex, of the referred to as *agonistic behaviour.*" Hinde points out that such a concept is often useful because of the presence of simultaneous tendencies to attack and to flee and associated occurrence of threat behaviour. Examination of Hinde's and other authors' uses of the term *agonistic* suggests that the temporal, functional, and possibly causal association of several diverse movement patterns is not the only reason for lumping them in a single category.

Although part of the usefulness of the term lies in its inclusiveness, part also is due to the difficulty encountered in dealing with the concept of aggression. In an excellent review of the topic, *Johnson* cites dozens of different examples of behaviours that could be said to have something to do with aggression. The social learning theorist *Bandura* also acknowledges the difficulty of defining the term *aggression*. His working definition, "behaviour that results in personal injury and in destruction of property," is qualified to extent that he wishes to make it clear that "aversive effects cannot serve as the sole defining characteristics of aggression"'; otherwise, the pain-producing activities of dentists or surgeons would be defined as aggressive. The term *agonistic* is essentially dismissed because it cannot be defined any more precisely than aggression. Wilson asserts that most aggressive acts serve as competitive techniques used in various ways to enhance success in sexual or resource competition in species exhibiting *contest* competition. Wilson's identification of the type of competition associated with aggressive activity is useful in helping to avoid confounding some of the many kinds of *competition* constructs with some of the more subtle types of *aggression* recognized by theorists. Wilson (1975) listed the forms of a aggressive behaviour as follows:

1. *Parental:* attacks on intruder when young are parent;
2. *Parent offspring*: disciplinary action by parent against offspring;
3. *Dominance*: control as a result of a previous encounter, of the behaviour of a conspecific,
4. *Sexual*: use of threats and physical punishment, usually by males of obtain and retain mates;
5. *Predatory*: act of predator, possibly including cannibalism;
6. *Antipredatory*: defensive attack by prey on predator, such as mobbing,
7. *Territorial:* exclusion of others from some physical space.

It seems fairly clear that while many scientists are largely interested in "*aggressive behaviour*," however defined, they find it useful to have a seemingly nonsubjective, largely operationally defined label to encompass may acts that occur during fighting or in hostile or competitive situations, especially when animal rather than human behaviour is considered. Thus, while some aspects of the operations, functions, and causes involved can be identified at all phylogenetic levels, the quantum leap associated with self-consciousness and purposiveness that occurs *demonstrably* only in man (and perhaps some other primates) prevents us from perceiving the kind of smooth conceptual transition that would justify development of a more or less unified model and definition of the system.

Situations Associated with Agonistic Behaviour

Agonistic behaviour includes all behaviour associated with the contest or struggle between individuals"—is definitely animal-oriented and emphasizes the situational component of this activity. Since agonistic behaviours are largely tool behaviours, functioning mainly in association with other organic or psychological needs, it is useful to examine the major contexts in which they are found. Agonistic behaviour tent to occur in situations involving conflict or competition or space, resources, mates or status, or for protection of self or young. Although *Wilson* suggests that most intraspecific aggression occurs in contest competition, some of his categories (parental disciplinary aggression, weaning aggression, moralistic aggression, predatory aggression, and aggression) are difficult to view as contests in the ecological sense that he utilizes (the other categories are territorial aggression, dominance aggression, and sexual aggression). Even though some of these types of behaviour occur mainly or exclusively in mammals, such a functional classification may be useful in identifying the adaptive circumstances associated with most agonistic activities.

A strictly functional view of aggression has some drawbacks, however, because it cannot encompass certain types of activities occurring in man that are often considered to be classic examples of aggression. *Johnson* cities examples like racial hatred, actions as difficult to categories consistently. The reason for this, of course, is that structures, physiological mechanisms, and behaviours may evolve under one set of evolutionary constraints and then relatively suddenly become exposed to an entirely different selective regime. The next step is often the appearance of altered use of the structures or behaviours. The mechanisms controlling aggressive activities in man undoubtedly evolved in precultural situations.

It seems highly unlikely that the same selective pressures remained operable in cultural man. The imposition of complex language, social, and economic systems must surely have altered the relationships between physiological causal mechanisms and environmentally linked overt behaviours. It has been seen that intraspecific fighting occurs mainly in the arthropods and vertebrates, and it is absent in species where it would have no survival, value. According to *Scott*, lack of adaptive value occurs in species in which the motor capacity for inflicting pain or injury is absent, and in which individuals are unable to recognize each other as individuals. Thus "Indiscriminate attacks on all members of the same species, including males, females, and young, could have only negative survival value." He feels that "defensive fighting in reaction to painful stimulation is the most basic type of social fighting," and that, "pan is the most basic primary eliciting stimulus ('releaser') for fighting."

Pointing out that such fighting can be observed even in the lower invertebrates, he concludes that social fighting may have evolved independently in various species from defensive, threat, or defensive attack behaviours. Now it is clear that the agonistic behaviour is highly situation-dependent, and while individual recognition may be an important concomitant in tetrapods, it has not been clearly demonstrated in many fishes.

Lack of evidence on this point need not impede the development of generalization, however, inasmuch as fishes are very clearly capable of discriminating among individuals exhibiting differences in sex, age, or other criteria that may identify a particular caste. Thus socially related agonistic behaviour in fishes often can be highly ordered to produce distinctive in fishes often can be highly ordered to produce distinctive social structures not necessarily dependent on recognition of all members of the group. Pair-bonding in some cichlid fishes, is of course, one example of individual recognition, but such examples are not common among fishes. Most agonistic situations occur among conspecific, and some scholars limit the concept to intraspecific encounters. *Lorenz*, for example, has stated that

> If we put together, in the same container, two sticklebacks, lizards, robins, rats, monkeys, or boys, who have not had any previous experience of each other, they will fight. If we do the same with two animals of different species, there will be peace—unless, of course, there is a prey-predator relationship between them.

Predator-prey relationships are considered to be distinctive by many scholars because, as *Hinde* puts it,

> In any one species predatory behaviour and intraspecific fighting are usually elicited by different external stimuli, depend on different internal states, usually involve some different movement patterns, and may involve different neural mechanisms.

Although *Hinde* admits that there may exist relationships between the two types of behaviour, many ethologists have supported a hard line for separating the two categories. It is interesting that *Wilson* utilizing functional criteria within an evolutionary framework, recognizes categories for predatory and antipredatory aggression in his classification scheme but does not discuss the motivational implications of this treatment. Given a strong a evolutionary-ecological bias, some difficult questions related to proximate causation or motivation can be evaded.

The Interpreting Causation

Huntingford, recently used uses the term aggression in a broad sense to cove behaviour ranging from over fighting to "supposed symptoms of boldness or confidence." *Huntingford* makes no simple distinction between "initiation of a fight, attack in the conventional human sense, and relationship or defense," though she admits that they may be distinct in their causes and consequences. She follows *Lorenz* in viewing conflict between animals in terms of three main situations: between members of the same species ("*social aggression*");

hunting and stalking as shown by a predator towards its prey ("predatory aggression"); and attack on a potential predatory by a prey species ("*antipredatory aggression*"). Her subsequent discussion suggests that analysis of the form of aggressive patterns provides limited information on the casual organization of the behaviour and that information on the context and patterning of aggressive actions is required before inferences on causation can be justified.

She points out that covariance in the frequency of occurrence of two patterns of behaviour in a constant environment over the same time span argues for the existence of common motivating or inhibiting factors, and she presents data showing that both overlapping and distinctive patterns of variation of aggressive behaviours may occur in a wide variety of carefully studied vertebrates. Thus, social aggression and predatory aggression (or antipredatory aggression) have been shown to increase (or decrease) together with variations in dominance, sex, isolation, and so on, it sticklebacks, mice, rats, canids, and primates, while the three types of aggression may also vary independently in rats, canids and fields. After pointing out that various physiological manipulations, such as castration, brain stimulation, or lesioning and ncurochcmical manipulation, may producc cithcr covariancc or independent fluctuation among the various patterns of aggressive behaviour, she argues that it is unrealistic to view the different kinds of aggressive behaviour as either irrevocably wedded or distinctly unique in causation.

She concludes that the evidence supports the view that "social, antipredator, and predatory aggression are neither invariably linked nor inevitably distinct motivationally" and then discusses in some detail the ecological factors that may result in covariance or divergence of aggressive systems. Huntingford's paper is a significant milestone in the understanding of aggression because it seeks to link ideas based on "casual analyses," often experimental in nature and sometimes buttressed by physiological investigations, with ideas from the increasingly important literature that identifies relationship between social organization and selection pressures imposed by specific ecological conditions. She argues convincingly that we cannot make strong inferences about the causal organization of agonistic behaviours without examining the selective pressures operating on populations of organisms. Knowledge of the ecological relevance of these behaviours will provide meaningful insights into the realm of potential selective pressures producing convergent or divergent patterns of behavioural control.

The validity of our inferences thus will depend to considerable extent on the soundness of our examinations of the ecological relationships within which each set of social behaviours evolved. *Huntingford* provides a scheme for evaluating the ways in which aggressive behaviours may have been integrated into the overall fabric of an animal's biology and for testing alternative hypotheses about behavioural control. It is worth noting that most recent published comments on the intraspecific nature of agonism have taken an either-or stance on the issue. For example, *King* omits species-specifically from the concept of agonistic behaviour "because or in interspecific competition are often the same as those patterns exhibited between conspecific rivals. "It seems possible that one's position on species-specificity may be determined partly or largely by the examples selective or the species or group most intensively studied by the scientist.

Each individual may thus be strongly influenced in his views by animals exhibiting either convergent or divergent mechanisms for control of *aggressive* patterns, and may thus be unlikely to perceive the ease with which an alternate mechanisms might evolve in species operating under distinctly different selective regimes. In attempting to assess the significance of similarities or differences in the situation eliciting agonistic behaviour in fishes as against the tetrapods, we must first acknowledge the enormous constraints imposed on the form and movement patterns of vertebrates by the medium in which they live. Fishes, living in a dense yet buoyant medium, typically utilize in locomotion a complex, powerful lateral musculature, controlled and toned by the action of paired and median fins. Spines and other armaments can be used as defensive weapons in some groups, but most fish can attack only by biting, butting, or ramming some part of the body against the opponent.

Conversely, most tetrapods have moved into a medium 800 times less dense than water and have been forced to develop a supportive appendicular skeleton, a largely nonmetameric parietal musculature, and a highly differentiated appendicular musculature in order to support the individual during locomotion or other activity. The addition of well-developed, operationally flexible appendages permits many tetrapods to increase the variety of movements that can be used in signaling attacking, and defending themselves. Despite this profound distinction, it appears that remarkable similarities exist in the contexts associated with agonistic behaviours in the two vertebrate types, and that analogous behaviours often occur in analogous situations.

There is a wide variety of visual fishes and tetrapods (including man) utilize autonomic responses a signals to convey rather precise information about motivation in fighting contexts (i.e., colour changes in fishes and reptiles, piloerection in birds and mammals). *Hostile* encounters is fishes often contain various combinations of autonomic and nonautonomic responses, such as fin-spreading, fin-flicking, opercle-spreading, tail-beating, mouth-gaping, and pendulum movements that seem to have analogs in behaviours like wing-raising, feather-flashing, gaping, pecking, necks-stretching and pendulum movements in birds and back-arching, rearing, tooth-baring, head-raising, tail-rattling, and foot-thumping in mammals. These behaviours and the less equivocal movements, such as biting, butting, slashing, grasping, kicking, pecking, wing-bearing, chasing, fleeing, freezing, and others, all occur in various combinations in situations in which individual priorities or needs must be determined. Many of the behaviours found in territorial fights, dominance encounters, and other competitive situations seem to involve movements that serve either to increase the apparent size of the actor or to expose specially coloured structures or potentially dangerous weapons. Conversely, behaviours occurring in submissive or subordinate individuals often tend to decrease structures. The widespread distribution of these behavioural phenomena among many vertebrate group leads one to ask whether or not such operational similarity is due to similarities (or perhaps even homologies) in the sensory, central, and neuromotor mechanisms of vertebrates at different phyletic levels.

Interspecific Territoriality

Agonistic behaviour occurs widely among non-colonial social animals and is prominent in most vertebrate groups. Among vertebrates, excluding man, it occurs most commonly, but not exclusively as a competitive technique used in acquiring actually or potentially limiting resources. Since different resource needs of a species may each requires the use of some kind of competitive technique (i.e., aggressive act), the species may evolve agonistic systems in which the overt acts are quite distinct, with distinct causal factors operating in each situation; or systems in which the overt acts are similar, but causal organization is distinct for each context, or systems in which both overt acts and causal factors are quite similar, even indistionetly different context. Perhaps less commonly similar causal mechanisms may be associated with overtly different acts is different functional contexts.

Most agonistic activities occur in predatory or antipredatory situations, or in fighting, usually in association with territorial or dominance encounters. *Hinde, Esser, Johnson, King, Stokes and Wilson* provide recent reviews documenting the functions of these important social phenomena and the ways in which agonistic behaviours are organized in time and space to achieve adaptive social goals. Although the concept of territoriality has been well known to vertebrate biologists since at least 1920, when *Howard* published his treatise on territoriality in bird life, the causes and functions of various types of space defense have been the subject of almost continuous debate among ecologists and ethologists attempting to develop a comprehensive model for the phenomenon. *Noble, Nice, Armstrong, Hinde, Tinbergen, Carpenter, Lack, Wynne-Edwards, Crook* and many others have attempted to examine the fundamental nature of territorial behaviour and the ultimate causes (adaptiveness) associated with it.

Despite the fact that interspecific territory defense was known as far back as Howard's original work, many scientists, have maintained that the territory construct was intimately linked to conspecific aggression. *Murray* recently argued that the interspecific territorial aggression of red-winged blackbirds toward tricolored blackbirds is a case of a misdirected intraspecific territoriality, and he attempted to show that such behaviour while selected against, could maintain itself under certain geographical and ecological conditions. However, others have suggested that interspecific attacks are normal and functional in the territorial activities of many birds. *Rasa* described a case of a teleost fish the Pacific damselfish *Pomacentrus jenkinsi* that defended a year-round territory against a variety of intruders.

The damselfishes (*Pomacentridae*) represent one of the most common groups of fishes found in tropical waters around the world. They are generally small, some are brightly coloured, and they inhabit a variety of niches (from the open-water, schooling planktivores (*Azurina*, *Chromis* sp.) through close association with the bottom (*Hypsypops*, *Microspathodon* and *Pomacentrus*) to commensalism with sea anemones (*Amphiprion*). Because of their interesting breeding habits and their occurrence in relatively shallow waters, they have been the subjects of more behavioural research than any other group of marine fishes. Most species studied have been show to defend breeding territories during the reproductive season. Among the more important of these studies are those of *Verwey* on *Amphiprion percula*, *Abel*, on *Chromis chromis*, *Turner* and *Ebert* on *Chromis punctipinnis*, *Stevenson* and *Dascyllus albisella*, *Helfrich* and *Abudefduf abdominalis, Limbaugh* on *Hypsipops rubicunda, Sale* on *Dascyllus aruanus, Rasa* on *Microspathodon chrysurus, Low* on *Pomacentrus flavicauda* and Myrberg on *Eupomacentrus partitus*. Of these species, only three, *H. rubicunda*, *P. flavicauda* and *P. jenkinsi* were known to defend their territories year-round.

Rasa had suggested that adults defended feeding territories against all fishes invading the territory space. However, *Clark* concluded that year-round territory defense in *H. rubicunda* was more closely associated with maintenance of an optimum spawning site, and thus its principal function was to increase reproductive success. Sometimes—territorial boundaries were relatively easy to discern but sometimes shifted slightly with topographic changes (loose coral or algal clumps may shift as a result of wave action) or movements into or out of the vicinity by other conspecifics. Most territories on Kaneohe reef sites had at least one clump of algae growing in them, and all had some kind of shelter hole(s) into which the fish disappeared when frightened. Fishes often were observed grazing on the algae and associated *Aufwuchs* that constituted most of their diet.

A reasonable hypothesis about the function of territorial behaviour in this species might be that such behaviour serves to protect a food source for the territory defender. If this were true, fishes receiving the brunt of most attack behaviour should be those that are also herbivoures or those that resemble food competitors physically or behaviourally. The most intense aggression should also be directed toward ultimate competitors—conspecifics. *Low* actually found such a situation in interspecific aggression in an Australian pomacentrid, *P. flavicauda*, which defended its territory against 38 species in 12 families, but ignored 16 other species in 6 families. All of the fishes attacked were food competitors (*herbivores*), while those not attacked were all *carnivores*. It has been observed that a common carnivorous wrass (*Thallasoma duperreyi*) that has almost never been attacked on Checker Reed was one of the prime targets on the second reef.

Although the reefs were only a few hundred yards apart, the density of fishes of the two species, and many others, was much greater on the second reef. A third set of observations were made on the reef face of Coconut Island. In the absence of an observer, many species rarely or never previously observed to enter. *P. jenkinsi* territories were found to be common, though transient, visitors. A number of these, including some other pomacentrid species, were primarily carnivorous and yet were frequently attacked by *P. jenkinsi*. One of the most surprising observations made was that very few conspecifics were attacked, even when they penetrated deep into to neighbour's territory. One might expect that a conspecific, being an ultimate competitor for both food and space, would elicit the greatest amount of threat and attack. This appears to be a good example of the *Dear Enemy* phenomenon discussed by *Wilson*. Because it is energetically wasteful for adjacent territory holders to fight constantly, they develop means for recognizing neighbours and reducing hostility toward them (through habituation or learning of individual traits).

Most previously known examples of the phenomenon are birds or mammals, and *P. jenkinsi* appears to be fairly unique among fishes in exhibiting this trait. A number of significant papers on the behaviour and biology of damselfishes have been published, and they point quite clearly to the nature of some key theoretical issues that can be clarified by the execution of additional work on *P. Jenkinsi, Smith* and *Tyler* noted that *Eupomacentrus planifrons* on a reef in the Virgin Islands defended discrete territories against conspecifics and some contra-specifics, and suggested that *E. partitus* was excluded from that portion of the reef containing *E. planifrons* territories. *Emery*, working on Alligator Reef in the Florida Keys, described territory defense, often apparently based on defense of feeding or shelter sites, in *Chromis insolatus* adults, *Chromis scotti* adults, *Eupomacentrus leuconstictus*, *E. partitus*, *E. planifrons*, and *E. variabilis*) in which territory defense is especially strong in juveniles); he suggested that the abundance of food and variable habitats on the reef permitted the damselfishes to develop fairly distinctive niches, with a consequent minimal competition for food or habitat.

Unfortunately, he provided little information on the permanence of territorial behaviour for most species and none on the kinds of fishes chased from territories. *Keenleyside* reported long-term territory defense in *Abudefduf zonatus* at Heron Island, Great Barrier Reef, and indicated that similar types of territorial patterns existed in *Pomacentrus flavicauda* and *P. tripunctatus*. He suggested that "A *zonatus* is a predominantly substrate species as adaptations to permanent residence on the shallow, inner reef flat." He also pointed out that *A. zonatus* was less aggressive toward conspecifics than toward small labrids, which are egg predators of many Heron Island pomacentrids. Unfortunately, he presented no quantitative data on attack frequency or species attacked, other than the above brief statement. Field observations and experiments on the territorial behaviour of *Eupomacentrus planifrons*, a drab-coloured Caribbean damselfish have been made recently by *Myrberg* and *Thresher* and *Thresher*. Using videotape analysis and live fish in bottles placed near the residence of territorial males, they concluded that these fish are capable of discriminating among a wide variety of species and reacting differently to them. This differential reactivity was formalized on the concept of serial territory, whereby a defending fish would swim out from its residence for greater or lesser distances, depending on the species approaching (or presented in a bottle) and the time of year.

In general, conspecifics were attacked farthest from the residence, congenerics somewhat closer, food competitors closer yet, and non-competitors closet. These species-species boundaries expanded during the reproductive season and contracted after spawning was over. Data on the number of nips directed at fish introduced (in bottles) directly in front of the residence also showed differences in attack patterns, but congenerics were attacked at about the same levels as other species, and conspecifics were attacked less than any other congeneric, *Myrberg* and *Thresher* suggested that "the territory appeared as a series of areas surrounding the residence, each area apparently reflecting the amount of space that secures, a limited resource(s) from members of a particular species," and that specifies eliciting more nips at the residence than conspecifics"—must have been considered as greater threats at the residence, while just the opposite was apparent for the areas surrounding the residence. This indicates that there is a difference in functional significance of these two features. (i.e., attack distance vs. number of nips at residence, RJM)." They go on to point out that protection of a good supply, maintenance of a nest site and protection of eggs, and maintenance of a residence may all be functions of territoriality in this species.

Thresher presented data on videotaped interactions of territorial males with other members of the fish biota and came to generally similar conclusions. Prior to 1969, most species were believed to act territorially (primarily toward conspecific) only during the breeding season, when breading adults defended the spawning site. We know now that many species defend territories for considerable periods outside the breeding season, and that attacks on contraspecifics are not cases of *mistaken identity*, as has been postulated by some scholars. Indeed, several studies have clearly demonstrated that territory defense in some damselfishes is a highly structured phenomenon with considerable discrimination involved in the determination of who will be attacked and how vigorous the attack will be.

Several authors have suggested that territoriality may function is defense of limited resources, such as food, spawning site, or shelter, or some combination of the there. Most species defending reef sites over long periods of time are herbivores or omnivores feeding on algae and other sessile organisms, a phenomenon indicating a high probability that a primary ultimate cause for permanent territoriality is the need to protect a food source. *Syrop* found that *Pomacentrus jenkinsi* maintained a higher standing crop of algae within territories that occurred in adjacent non-defended areas. Furthermore, territory size varied both with the productivity of the area and the abundance of herbivorous intruders in the vicinity.

Another ultimate cause (function) probably lies in the protection of a spawning site, since *Myrberg* and *Thresher* clearly showed enlargement of serial territory dimensions during the breeding season. Finally, since these species utilize shelter holes in the reef at night and during the day when threatened, maintenance of a specific shelter site could serve as another ultimate cause. *Smith* and *Tyler, Collette* and *Talbot* and *Hobson* all present data demonstrating the important of shelter areas to many reef-inhabiting fishes. Although the studies of *Myrberg* and *Thresher* and *Thresher* have added new dimensions to our knowledge of operational aspects of the territorial behaviour of pomacentrids, their speculations on causation must be critically examined. For example, the former state that.

> If the entire area of defense is based upon a single limiting resource (e.g., a specific type of food, such as benthic algae), the important variable across species lines, should in most cases be the amount each threatens that resource. If, in contrast, the different sized territories reflect the fact that various resources are being secured by the resident, then certain boundaries secure specific levels of one resource, while other boundaries secure specific levels of others. This is perhaps the more likely possibility.

Likewise, *Thresher* states that

> Territoriality by *E. planifrons*, therefore, is probably multifunctional. The data suggest that it serves to protect a food supply, the spawn, and the residence. Additional functions especially those related strictly to intraspecific defense, can only be speculated upon. The size of the territory maintained against any given intruder is a function of the degree with which it threatens any or all these all of these resources vital to the threespot.

Although these authors are probably, correct about the functions of territoriality in *E. planifrons*, the relationship between function (ultimate cause) and (proximate) causation becomes so intimate in the discussion that the two appear to be inseparable. Thus, in gets the impression that a certain species is attacked farther from the residence because it is a greater food competitor than others in the community. This apparent confounding of levels of organization may be misleading, because attack behaviour patterns could depend upon patterns of proximate causality quite unrelated to the degree of competition or threat to a resource offered by a given species. This is almost a truism in the behaviour of lower vertebrates, where elaborate systems of releasers or other control and communication mechanisms typically serve to integrate limited sets of motor patterns that combine in various ways in different functional contexts.

Mechanisms controlling responses to external cues (body shape, size, activity patterns) or regulating internal thresholds (habituation, diel activity rhythms, arousal processes) have been programmed to provide the animal with a motivational and motor capacity serving the ultimate cause (resource protection) without necessarily precluding use of some of the motor components in other contexts having other functions. To hypothesize a mechanism whereby proximate and ultimate causes are nearly identical is to require either a cognitive capacity for these animals far greater than is known to exist in fishes, or else a programmed behavioural template so complex that it is difficult to envision. Thresher tested reactions of *E. planifrons* to *Holacanthus tricolor* by presenting the latter in bottles that could eliminated visual, olfactory, or acoustic cues. He also studied attack distances in relation to normal movements of four intruding (in bottles) species, including conspecifics, and tested for the effects of form and color by introducing to territorial males seven species of hamlets (*Hypoplectrus*), three of which are colour mimics of three damselfish species also used in the experiment.

He concluded that interspecific attack is mediated by form recognition, while conspecific attack depends on response to booth form and colour. He discarded the idea that interspecific attacks were based on *mistaken identify* and suggested that there were distinct differences between the displays directed to conspecifics and closely related congeners and those directed to non-relatives. In the case of *Pomacentrus. jenkinsi*, where many territory defenders are packed into a limited area, and large numbers of a wide variety of species are almost constantly moving over the face of the reef, a rather different motivational organization may have evolved. This social organization would guarantee, in terms of population, the protection of critical resource(s) and thereby the maintenance of adaptive genomes without necessarily fixing agonistic behaviours along rigid stimulus-specific lines.

For example, if the species is programmed to respond with increased patrolling and chasing activity to higher densities of fish in the vicinity, non-competitors) could be attacked when a defender's attack threshold was lowered greatly by large numbers of fish in the vicinity (perhaps a general arousal phenomenon). In an overall sense, this could be adaptive if the (food) energy saved by attacking competitors at high rates when fish density is high (but still below the level producing attack of non-competitors) was greater than that expended in attacking non-competitors at extremely high densities. By cuing the threshold of attack to a non-specific sensory input, some difficult problems of behavioural control might be solved.

Precise measurements of the energetics of this situation would be extremely difficult to obtain, but experimental verification and quantification of these phenomena would provide at least circumstantial evidence that a density-dependent factor is involved in the control of attack behaviour. *Grant* pointed out that less work has been done on interspecific competition and territoriality in mammals than in birds, but he predicts it should be more widespread among rodents than among birds, because rodents have not been able to achieve niche separation in the vertical plane (as is so common in birds), and their predominant utilization of the structurally simpler horizontal dimensions of the environment should lead to more contact and increased possibility for competition.

Wilson provided a brief but comprehensive review of the subject in which he utilized elements of both Orians and *Wilson's* and *Murray's* ideas to create an interesting model for social evolution. Although his theory is too lengthy to pursue in detail here, a major conclusion that he presents is that interspecific territoriality is most likely to occur in cases where two cognate forms have recently evolved from a common ancestor and have just come into contact with one another. This argument leans somewhat toward Murray's ideas but shares an important feature with most of the published papers on the topic interspecific territoriality is largely a mutual phenomenon, with two species actively competing with each other and both forms exhibiting territorial behaviour. Most of the avian work has been conducted on passerine species exhibiting reproductive territoriality.

Although food and other resources are undoubtedly critical factors in the evolution of territorial behaviour in birds, the entire system is in a different stage of development than that found in benthic damselfishes. The tropical reef ecosystem is an ancient one, and the fish assemblage is staggeringly diverse. Fishes are notoriously opportunistic feeders, and though one an find considerable niche separation within a family, the sheer numbers of different families and genera occurring on the reef are certain to produce numerous instances of niche overlap-even among fairly distantly related species. Thus, for food-based territoriality to be adaptive on the reef, it simply cannot be the case in many birds and mammals. The combination of temporal and community differences seem to have brought about territorial systems based on rather different functional principles involving distinctly different causal systems.

Causation

It has recently been pointed out by *Wilson* that the social organization results from the competitive action of two broad classes of phenomena; *phylogenetic inertia* (factors tending to limit evolutionary change) and *ecological pressure* (factors promoting change via natural selection). Although many recent authors have attempted to examine social activities (including agonistic behaviour) within the framework of such an evolutionary model, the great bulk of the literature does not yet reflect the importance of acknowledging and identifying the contributions of these two classes of competing pressures in the development and maintenance of specific social patterns among vertebrate groups. *Wilson* himself tends to slide over the specific phenomena responsible for producing particular social patterns when, in his desire to develop a methodology leading to a "unified science of sociobiology" he states that he has become "increasingly impressed with the functional similarities between invertebrate and vertebrate societies and less so with the structural differences that seem, at first glance, to constitute such an immense gulf between them."

The specific parameters and quantitative models that would serve in the development of the kind of overview that he seeks must necessarily depend upon a considerable degree of simplification, generalization, and abstraction that would eliminate precise definition of the details of the evolutionary process as reflected by the specific structures and behaviours exhibited in each distinctive phyletic line. Any attempt to deal with diverse behavioural phenomena without examining *both* proximate and ultimate factors as they are reflected in the integrity of the organism can only provide us with a biased picture of the multidimensional nature of agonistic systems and other social phenomena. The preceding discussion of interspecific territoriality illustrates the difficulty of interpreting the behaviours involved without having some insights into both their function and the nature of the stimulus-response mechanisms utilized. Since our knowledge is still imperfect, an integrated approach to the study of these phenomena offers a real opportunity to understand better how they are controlled and why they exist in somewhat different forms among different species and groups.

Many people feel that the key to understanding the causation of agonistic behaviour lies in understanding the nature of physiological mechanisms associated with aggressive acts. This area has become one of the fatest-growing and most productive fields of ethological-neurological-endocrine research, and it offers great promise for clarifying some of the more complex behavioural puzzles that ethologists have faced over the last 30 years. Despite the enormous literature on physiological orrelates of agonistic behaviour, it is difficult to formulate well-founded generalizations, partly because most of the work has been done on mammals but also partly because of the high degree of specialization and differentiation found among vertebrate species. One of the best-known examples of vertebrate sensory specialization (though it is related to feeding rather than fighting) occurs in the frog, where the need to detect and capture mobile insects has led to the development of a unique *bug-detecting* system in the afferent visual apparatus. The absence of similar specializations in other vertebrates that have been studied may be due to changes in the extent of peripheral processing of visual stimuli that occur in concert with the shift from small-brained to large-brained vertebrates.

It appears that the highly complex synaptic interconnections found in the retina of lower vertebrates are necessary to accommodate considerable peripheral data processing in groups with poorly developed central mechanisms. The highly sophisticated visual cortical mechanisms of mammals, however, permit a simplification of peripheral synaptic organization. *Capranica* and *Frishkopf et al.* demonstrated a correlation between threshold sensitivity of auditory nerve neurons and spectral characteristics of bullfrog calls, but some central processing appears necessary in the elicitation of appropriate responses. The fact that most birds and mammals appear capable of using visual, acoustic and/ or olfactory signals for recognizing *individuals* in social groups suggests that much more central processing of sensory data occurs in amniotes. In the medulla of many fishes there are large Mauthner cells having the ability to initiate speedy locomotion (Tail-flip) by linking acoustico-lateralis, optic, and cerebellar systems with motor neurons in the lateral musculature via collaterals from giant axons in the spinal cord, providing a reflex escape mechanism invaluable to the teleosts and tailed amphibians that posses it. Why should the Mauthner system by present in most teleosts and tailed amphibians, yet absent in sharks and terrestrial vertebrates? Its absence in sharks in mainly an issue for the tetrapods.

The most parsimonious way to modify a primitive Central Nervous System (CNS) to activate rapidly a metameric locomotory musculature is to develop a reflex system utilizing giant axons carrying stimuli quickly to the muscle masses. In an evolutionary sense, it would be impossible to evolve or utilize a mauthner-type apparatus in tetrapods simply because of the loss of metamerization and the development of a more highly differentiated neuromuscular coordination system. Furthermore, the increased complexity and organization of the CNS of higher vertebrates permits development of a wider diversity of neurobehavioural mechanisms for coping with novel or threatening stimuli. One kind of reflexive mechanism that can produce quick response of the whole organism in higher vertebrates is the orienting response studied by *Sokolov* and others. It is associated with orientation and alertness rather than with the startle response.

Subsequent responses to the stimuli eliciting orientation depend upon the consequences of initial reactions to the stimuli, but are suggested by Sokolov to be dependent on cortical process. One can find examples of both sensory and neuromotor mechanisms that are organized in fundamentally distinctive ways at different ends of the vertebrate phylogenetic spectrum. From the perspective, the development of adaptive structural and functional complexes must be viewed as a mosaic of evolutionary trajectories, each represented by species or groups moving into new niches of varying breath. Once initiated, these trajectories operate to constrain each viable phylogenetic unit, though the nature of these constraints would very with time and the maturity of the ecosystem within which any unit is evolving.

In some cases, once a particular trajectory had passed a given threshold (determined by genotypic modifications), further change in structure or function would be appreciably circumscribed, and the system would assume a determinate nature. Many of the specializations therefore, must be considered to be consequences of the phylogenetic process, dependent upon changes dictated by the needs of animals penetrating different worlds and developing new modes of structure, function, and behaviour that permit them to cope with novel problems.

Although each new major adaptive mode has produced a spectrum of characteristically unique structures and behaviours, these modifications were all based on the general vertebrate plan, and each represented new ways of organizing available materials. Quantum jumps to distinctly different levels of organization occurred rarely, as they may have in the case of the evolution of the human brain and culture, where certain emergent properties may exist. The *apparently progressive* nature of many of these changes perhaps can be illustrated by a few brief examples. The visual projection area in fishes lies in the *optic tectum*, consisting of two bilateral dorsal mesencephalic hemispheres. The tectum contains at least six layer and serves and the major integration centre for visual and other sensory inputs and for information ascending or descending from other neural centers. However, while the tectum serves as the highest information and integration center for fishes, the forebrain and tectum share these functions about equally in reptiles, the basal ganglia take over most such functions in birds, and the *neopallium* is the major integrative structure in mammals.

The tectum (*superior colliculus*) of mammals serves mainly as a visual reflex center and as a relay and transducing gateway to the temporal cortex, which is the site of the visual projection. Like wise, the endocrine systems of vertebrates exhibit a strong morphological tendency to form distinct vascularized organs in higher groups, whereas fishes still retain diffuse, often scattered, endocrine cells (i.e., adrenal cortex and medulla, and thyroid tissues) not far advanced structurally from their presumed ancestral homologues. Presumably this shift reflects an evolutionary trajectory influenced by two major forces, according to *Hoar*: "The need for the seasonal regulation of reproduction, and the abandonment of non-selective filter feeding for the life of a selective, and often predaceous, feeder with major problems of regulating digestive and metabolic functions." *Moyer* has attempted to incorporate many of the often diverse observations on neural and endocrine substrates of aggression into a scheme recognizing eight different functional categories. Each category is based on some complex of neural, endocrine, and contextual characteristics that serve to identify the distinctive type of aggression though both *Wilson* and *Huntingford* criticize Moyer's classification because the categories are *introspective* and are not based on any consistent criteria, his contribution has been valuable in pointing out the diversity of control mechanisms underlying *aggression* in mammals.

It has become more difficult to view *aggression* as a unitary construct in recent years. Thus, while neural and endocrine substrates for the control and development of structure and behaviour show a degree of continuity throughout the vertebrate line, shifts in the localization of functional elements and changes in the magnitude of influence that a particular neural or endocrine mechanism exerts on behaviour may be quite prominent. One can find numerous examples of distinctive physiological mechanisms underlying the sensory, neural, and motor function involved in many behaviours, including agonistic activities, occurring in different vertebrate groups.

Yet similar stimulus properties often elicit similar kinds of behaviours, considering the morphological equipment available. This similarity is due to the fact that there is a relatively restricted number of social problems that must be dealt with in any species, and that there are a limited number of ways in which any vertebrate kind can cope with them. To borrow another well-documented example from morphology, consider the tapetum lucidum, a reflective tissue found in the retinae of many nocturnal animals. According to *Walls* and others the tapetum may be composed of guanine (fish and many other vertebrates), shiny tracheal tubes (insects), tryglyceride spheres (some fishes), white collagenous fibers (musk-ox), or riboflavin crystals (bush baby, *Galago* and *garpike*).

Quite obviously, the adoption of nocturnal habits increase the probability of development of a structure facilitating dim-light vision, and the specific chemical or physical mode utilized is almost irrelevant, in terms of adaptiveness. Examples, such as the preceding, of groups or species acquiring morphological or behavioural adaptations for a mode of existence strikingly different from that of other members of the same phyletic line are readily identified and require no further discussion here. However, subtle changes in fairly closely related groups may not be recognized at all as representing convergence; thus assumptions about the uniformity of causal mechanisms producing similar overt behaviours in different groups may not be justified where the entire chain of causality has not been studied. There are many basic coordinating and integrating mechanisms persisting throughout the vertebrate line. The ways in which different species and groups operate to regulate the division of, or competition for, limited or limiting resources—that is, food, space, cover mates, young, and so on. Thus, some response patterns, such as withdrawal from larger or approaching stimuli, may reflect the operation of homologous sensory and/or central mechanisms that are evolutionarily conservative in that they have maintained an association with adaptive responses throughout vertebrate phylogeny.

Other reactions, similar in apparent function and context, may depend on quite distinctive physiological causal mechanisms that evolve in response to novel adaptive necessities. A mosaic of conservative (*homologous*) and specialized (*often convergent*) mechanisms are available to most vertebrates for the elicitation and control of agonistic and other social activities. The relative contributions of these mechanisms are probably determined largely by the nature of the species and the adaptive context in which the interaction occurs, and analysis of the causal organization of behaviours must be carried out in each group of species with comprehension of the fact that parallel contexts and behaviour need not imply parallel causality in its fullest sense.

It is tempting to focus attention on the adaptiveness and survival value of agonistic behaviours because of the breadth of variation extant in causal mechanisms subserving overt activity, there has been too much confounding of functional operations, and causal constructs. These conclusions may seem self-evident, but much of the controversy about the nature of aggressive motivation may be due to the fact that this duality of motivational origins, and the multidimensionality of species-typical agonistic causality, have seldom been examined in light of a truly broad evolutionary perspective.

The structural, functional, and behavioural attributes in any animal species in any phyletic line represent a composite, or mosaic, of features derived from ancestral types and developed in response to specific selective pressures of the environment and life-mode (niche) of that particular population or species. The degree of singularity or novelty in the overall organization of the animal kin d depend largely on the uniqueness of the specific niche occupied by that animal. Two factors tend to operate to preserve archaic or conservative features of an organism's design: (1) drastic, rapid changes often tend to be maladaptive; and (2) adaptations that successfully deal with major problems faced by a species may be retained by descendant forms as long as the same types of selective pressures are exerted on successive evolutionary types. The fact that all vertebrates have not developed identical ways of utilizing this basic substrate to organize their competitive activities simply reflects the extent of adaptive radiation in the phylum. Depending on one's scientific goals, one can emphasize the similarity found among adaptive strategies that deal with similar evolutionary problems, or one can focus on the specific differences among related species that make them seem so unique.

6

COMMUNICATION

Many of the examples discussed throughout this book have involve communication between animals, for this has always been a subject of particular interest to ethologist. Indeed the theories they developed were in some ways biassed by this interest. Acts of communication do tend to be very fixed in form, to act as releasers for other actions, and to develop very similarly in all members of a species. Perhaps ideas such as that of the fixed action pattern or the innate releasing mechanism would have come less easily to mind if ethologists had been more interested in grooming or feeding behaviour than in aggressive or courtship displays. But then they had every reason to be enthusiastic about studying the latter, because the signals that animals produce in communication are often remarkably striking. Discovering just what they are saying to each other is one of the most fascinating fields of ethology. Whenever, we see an animal that is brightly coloured or boldly patterned, or one that is making a loud and obvious noise, we can be pretty certain that communication of some sort is taking place. Standing out from the background is dangerous and there must be some benefit to doing so which offsets the risks involved.

Most often the communication is between members of the same species, and the risks come when predators cue in on this as a means of finding a meal. Sometimes, however, different species communicate with one another and here it is often prey animals that forsake their camouflage and communicate directly with predators that might eat them. These different sorts of relationship between predators and their prey make a useful starting point as they illustrate some basic principles of animal communication.

How Signals Convey Information

Discrete and Graded Signals

Some signals are discrete (digital) and others are graded (analog) The alarm calls of many species, which are typically given at the same intensity each time and are relatively constant across species, permit communication among different species. These discrete signals are of a frequency and duration that make them difficult for predators to pinpoint. Graded signals may vary in intensity as a function of the strength of the stimulus.

The waggle dance of the honeybee illustrate both the complexity and the graded properties of signals. The number of turns per minute is inversely proportional to the distance from the food source, and the duration and liveliness of the dance increase with the quality of the food. Signals that might at first seem discrete often, upon closer study, turn out to be graded. Although the bursts of light emitted by fireflies seem to be discrete and species-specific, they vary in intensity and duration under different conditions.

Distance and Duration

Although most species are limited to about twenty to forty different displays, signals can vary in several other ways that increase their information content. The distance a signal travels may vary. In some cases, the detection threshold of the receiver is no more than several molecules of a substance, as with sex attractants of many insects. Thus a small amount of material produced by a female can be detected by a male several kilometer downwind.Visual displays, at the other extreme, usually operate over much shorter distance. The duration of a signal may also vary. Alarm signals, such as chemicals produced by many invertebrates, may have a localized, short-term effect and thus a rapid fade-out time. Signals such as male secondary sexual characteristics in polygynous species can last for long periods of time. Information about reproductive status, like the antlers of male deer or the bright red perineal region of female rhesus monkeys in enstrus, is conveyed during breeding during season. In contrast, the brightly coloured epaulets on the red-winged blackbird, although present during the entire breeding season, are conspicuous only when the male exposes and erects these feathers during the song spread.

Composite Signals, Syntax, and Context

Two or more signals can be combined to form a composite signal with a new meaning. For example, equids such as zebras communicate

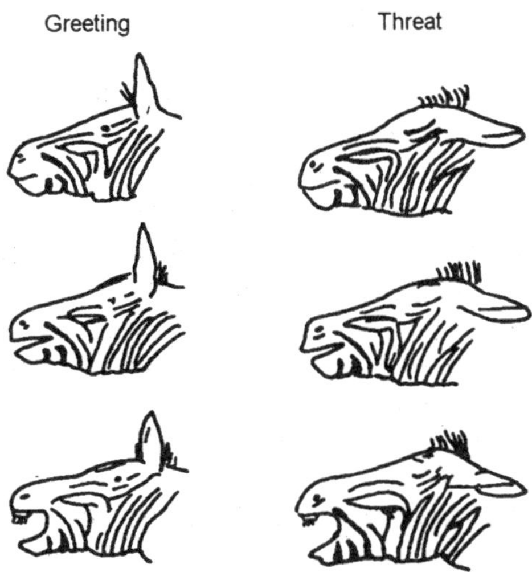

Fig. 6.1. Composite facial signals in zebras. Ears convey a discrete signal. They are either laid back as a threat or pointed upward as a greeting. The mouth conveys a graded signal and opens variably to indicate the degree of hostility or friendliness.

hostile behaviour by flattening their ears and communicate friendliness by raising their ears (discrete signals). They indicate the intensity of either emotion by the degree to which the mouth opens (graded signal). The mouth-opening pattern is the same for both hostile and friendly behaviour. Animals can convey additional information with a limited number of displays through changing the *syntax*—the sequence of displays. For example, the two composite signal A and B would have different meanings depending on whether A or B came first.

No evidence exists that nonhuman animals use syntax naturally. However, in the rather special case of language learning in chimpanzees, chimps assemble words in novel ways as they communicate with their human companions. The same signals can have different meanings depending on the context-that is, on what other stimuli are impinging on the receiver. For example, the lion's roar can function as a spacing device for neighbouring prides, as an aggressive display in fights between males, or as a means of maintaining contact among pride members. The "ruff-out" display of the male cowbird serves in courtship as well as in conflict with other males.

METACOMMUNICATION

Increasing the information content of displays by *meta-communication*—communication about communication—is theoretically possible; one display changes the meaning of those that follow. We can see good examples in play behaviour; animals use aggressive, sexual, and other displays in play, but they precede such behaviour by an act that communicates the message that "what follows is play, join in". Canids such as dogs and wolves precede play with the play bow. Monkeys communicate play behaviour through a relaxed, open-mouthed face. Disagreement exists about the importance of metacommunication, and some observers have used the word a bit loosely. For example, male rhesus monkeys (*Macaca mulatta*) sometimes communicate dominance by carrying the tail elevated in as "S" shape over the back. While this posture conveys status and mood, it does not really change the meaning of behaviours that follows; rather, it communicates that aggressive behaviour is likely to follow.

Information and Manipulation

In the past, ethologists often used to refer to communication between animals of the same species as involving the sharing of information for their mutual benefit. Indeed, much communication may be just this. For example, when potential mates meet it is advantageous to both of them that they communicate to each other accurately information about the species to which they belong, for if they are not members of the same species both of them will waste their breeding effort. Strictly speaking, however, animals will produce signals when this is to their advantage, regardless of whether other benefits from receiving them. The examples in the last section showed this well, for prey animals send signals to predators when they gain by doing so and the gain they make is usually at the predator's expense. Exactly the same rule applies within a species, as we might expect from the idea that animals are essentially selfish, behaving so as to maximise their own inclusive fitness.

So, while the information that passes from one to another when a signal is transmitted may be useful to the one that receives it, this will not necessarily be the case, and there are some very good examples where it is better to think of animals as manipulating each other rather than sharing information. One of these is in the mating behaviour of the bluegill sunfish. Most males of this species do not mature until they are seven or eight years old, when hey set up territories to which they attract females or mating. The females are smaller and look quite different. Sometimes two of them may spawn simultaneously with one of these males, sperm and eggs being shed into the water as they swim around together on his territory. Deceit enters into this scheme of things because some males, around 20%, mature at only 2-3 years old, when much smaller than the norm. These look just like females and join in the mating of spawning pairs.

The territorial male does not drive such a rival off as he cannot tell that he is not a second mate, but actually he is a male and he shares in the fertilization of the eggs that the female produces. Of course, all males cannot mature early, or there would be no territorial ones to attract females in the first place, so this is an example of a mixed evolutionarily stable strategy, with a balance between the two possible modes of behaviour. In this example males mimic females and the other males, unable to see through the deception, are manipulated by the mimics. The case of the paid flycatcher mentioned in the last chapter can be put in very similar terms. Here the female cannot tell if a male is mated or not and so cannot choose an unmated one who will help her rear her chicks rather the one who will leave her to return to his first mate. Deception is possible in these cases because there are no cues that a signal receiver can use to discriminate between the classes of individuals involved, in other words to tell young male sunfish from females and mated flycatchers from unmated ones.

In other cases, however, deceit is impossible because there are ways in which other animals can see through it. Some good examples here come from cases where it pays animals to appear to be larger than they really are. This is often so in aggressive displays for, the larger an animal appears to be, the more a rival is likely to be intimidated into retreating without a scrap. It is probably for this reason that Siamese fighting fish extend all their fins as far as possible when swimming alongside an opponent and raise their gill covers when facing one. Both actions make them as large and striking as they can be. If only some fish displayed like this, as was probably originally the case, they would be likely to have deceived other of the same size into retreat. Thus appearing bigger would have to their advantage and natural selection would have made it spread through the population until, eventually, all fighting fish looked as large as they possibly could when displaying.

At this stage, however, the trick breaks down for each just looks an exaggerated version of its own size. If every fish can increase its apparent size by 20%, all will look that much bigger, but their sizes relative to each other will remain the same. Their opponents will gain accurate information about their size rather then being misled in

Fig. 6.2. Three bluegill sunfishes spawning. The large animal at the back is a full-growth adult male. Through they look similar, only the nearer of the two small individuals is a female.

any way. Another nice example is in the calling of male toads studied by *Nick Davies* and *Tim Holliday*. Male toads cling onto ripe females in a posture called amplexus which ensures that they fertilize the eggs as they emerge. Males often fight to take up this position and the male already in amplexus kicks and croaks to repel boarders. Larger males generally win in such encounters, as one might expect from their greater strength.

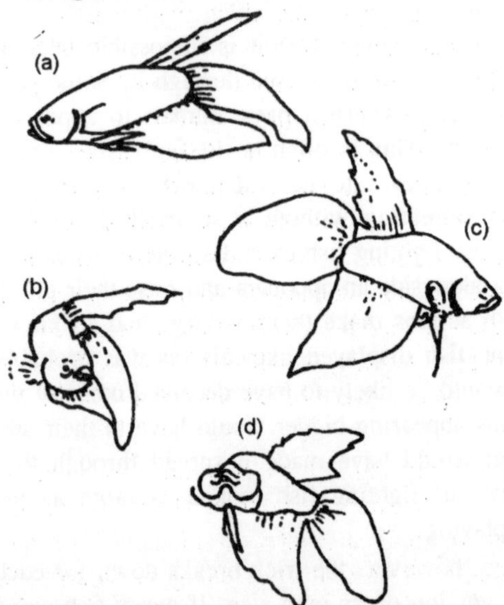

Fig. 6.3. The male Siamese fighting fish is a fine example of an animal that looks larger when displaying (c and d) then it does normally (a and b). The displaying fish raises all its fins when broadside to its rival (c) and its gill covers when facing it (d), thus making itself appear as large as possible.

Being larger, however, they are also able to produce deeper sounds, and Davies and Halliday showed that these too had an effect. The silenced males in by amplexus fitting elastic bands round their arms and through their mouths. They then saw whether other males attempted to displace them. As expected, medium-sized males spent more time attempting to displace small males than large ones. However, if the deep croaks of a large male were played from a loudspeaker, a small male in amplexus was harassed much less by his larger rival. Clearly toads can use the calls to assess the size of others and will be less likely to fight with those whose calls are deeper. Here again then, the information transmitted is accurate and the animal receiving it acts on it appropriately. But why do small toads not cheat, developing deeper voices and so appearing larger then they really are? The answer is probably that the larger an animal the deeper the sound it is capable of producing.

So, just as with the size of fighting fish, if all toads call as deeply as they can, the larger ones will still all more deeply than the smaller ones, and depth of croak will be an accurate reflection of size. To summarise, communication involves the transmission of information, in the form of a signal, from on individual to another. As

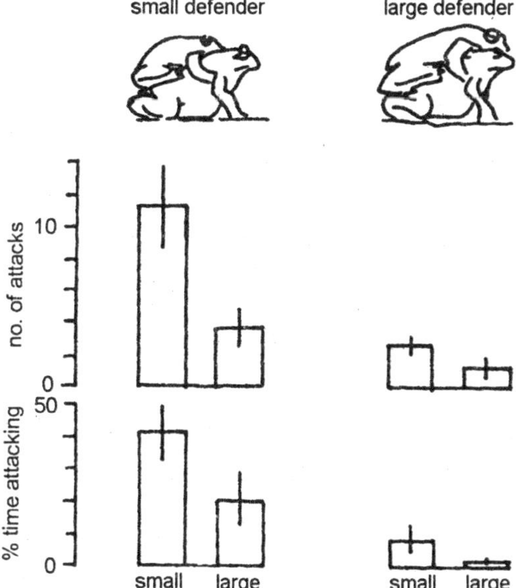

Fig. 6.4. Male common toads try to displace each other from the amplexus position on the backs of females.

we have discovered, this 'information' may be true or false, the important point being that producing the signal should be of benefit to the animal that does so. Sometimes deception may be possible, and the signaller may gain from it, but this will not be possible if the animal receiving the signal can detect that the signal is misleading. Deception tends, therefore, to be rather rare, for the more animals practice it, the more it will be best for the recipient to assume that the information it is receiving is false. If almost every plover which appears to have a broken wings is really feigning injury, then a fox discovering one would gain most by not following the adult but searching the area for a nest or chicks.

Messages and Their Meanings

A useful way of analysing communication was devised by the American ethologist W. John Smith. He pointed out that it could be looked at either from the point of view of the animal that sent the signal, in which case one would ask what the *message* encoded in it was, or from the view point of recipients, to see what its *meaning* was for them. Let us return to or singing bird. What is the message incorporated in the song and what meanings may it have for those that are listening? The most obvious point is that, as every bird-watcher knows, song is usually clearly distinct between species, so that it includes information as to the species to which the singer belongs.

Furthermore, song tends only to be produced by male birds in breeding condition, they usually only sing when they have a territory and in some cases, they stop singing as soon as they obtain a mate. In such a species the basic message of song may be 'I am an unmated adult male sedge warbler on my territory and in breeding condition.' In addition, there may be many subtleties incorporated in the message. Signals often vary from place to place, forming dialects like those in human language. Individual animals may also have idiosyncrasies in the form of the signal that they employ. They may have a range of different signals which convey subtly different information.

The singing bird may thus also be saying where he comes from, exactly who he is and perhaps even something about his motivation if the songs differ according to whether they are directed at rivals or at potential mates. Many animal signals are, like song, primarily concerned with reproduction, attracting mates and repelling rivals being their main role in communication. However, especially in species that live in social groups, there are many other messages that it may be useful to convey. Animals that move around in groups often have obvious patterns on them which can be seen some distance off and enable them to maintain contact with their companions. Call notes can serve a similar function, especially where visibility is poor. These are simple sounds and patterns, and they tend not vary much. Their messages may simply be 'I am here', though usually they also communicate the signaller's species, and there are cases where, despite their simplicity, they vary enough to indicate individual identity. The alarm calls referred to earlier are an example of another type of message, individuals in this case warning their companions of danger. The 'seeet' call of small birds is most reliably produced by the appearance of a hawk but, in the breeding season, birds with young may also call in this way when alarmed by a dog or a human near to their nest.

Most animal signals are like this: the message they convey is a generalised one rather than being precise like a word in our language. A famous exception is the alarm calling of vervet monkeys. These animals have several different calls which they produce when alarmed and three of these are specific to particular sorts of predator: a snake call, a leopard call and an eagle call. Here the information in the call is very precise: the calling animal might as well be shouting 'snake', 'leopard' or 'eagle', provided that its listeners were able to recognise the meaning of those words as we can do. What do its companions do when they hear one of these calls? In other words, what *meaning* does the call have for them. The answer is that this too is very precisely defined. Animals hearing the leopard call rush up into trees, they respond to the eagle call by running down to the ground and hiding in thickets, and the snake call leads them to approach and look down. We cannot tell if they have a mental image of the predator in question, but they certainly behave as if they were aware of the particular threat it poses. To understand messages, we have to study the particular situation in which an animal produces a signal, in other words try to discover what causes it to do so.

In the case of meanings it is the recipient of the signal that we must study: how does it respond when the signal is received? This response may vary from one animal to another. In the case of a contact call, for example, other members of the flock may approach so that the group keeps together, animals form other flocks may keep their distance, predators may be attracted with the prospect of a meal, and animals of other species are likely to ignore it as irrelevant to their interests. The same signal can thus have many different meanings, depending on the listener and on the exact context in which it is received. To stress a point made earlier: the critical factor is that,

Fig. 6.5. The response of vervet monkeys to three different alarm calls. (a) That produced in the presence of a leopard leads them to run into trees, (b) the eagle call leads them to run down into the undergrowth and (c) the snake call causes them to look don and approach.

for a signal to be produced, the advantages to the signaller must outweigh the disadvantages. Otherwise, there is no reason why evolution would favour signalling. There are thus many different messages that animals signals may convey, though it is unusual for these to be very precise.

Most often they provide rather general information about the motivational state of the signaller, through their features can also incorporate more specific information, such as the species or identity of the caller. The number of distinct signals that a species uses in not often larger. What makes the signalling system flexible and varied is that the same signal may mean different things to different recipients and the same animal may respond to it in different ways depending on the context in which it is received. Thus, with a comparatively small repertoire of rather stereotyped signals, animals can convey a wealth of information from one to another.

Signals

The message is what a signal says about its sender, the meaning is what the receiver extracts from it. Neither of these is tangible: they cannot be studied directly but must be inferred form behaviour. On the other hand, the signal is the physical form in which the message is embodied for its transmission through the environment.

For most animal signals, the form of the signal has no clear relationship to the message tat it encodes, just as the word 'dog' does not look or sound anything like the animal to which it refers. However, signals are not absolutely arbitrary. The modality in which they are transmitted and the form of the signal within that modality may both be determined by considerations of how the signal is most effectively transmitted given its function and the need to avoid the attraction of predators. A well-known example of this is the 'seep' alarm call of passerine birds, found in very similar form in many species and which even has equivalents in some mammals. Thin, high-pitched whistles like these have probably evolved because the sources of signals of this form are especially hard to locate and this is important in a call produced when a predator is present. Conspecifics do not need to know where the signaller is in order to react appropriately, and it is essential to the caller that the chances of the predator locating it mare minimised. Hence the calls of may species have converged on the same signal form, and these species will react to each other's signal, without it necessarily being advantageous for an animal of one species to tell those of another about the danger.

Fig. 6.6. Among Darwin's many interests was the expression of emotion by animals. He used anthropomorphic terms such as those given below, but today animal psychologists would avoid them in describing the behaviours shown.

There are other cases in which the form of a signal seems to be related to the message that it conveys. *Morton* (1977) argues that this is so with many bird and mammal sound, harsh, low-pitched calls being used in hostile contexts and purer, higher pitched one where the situation is friendly. He suggests that the former may be intimidating because harsh sounds of low frequency can only be produced by large individuals. Thus animals might be expected to avoid those producing deeper sounds than themselves, as size is such an important factor in success in fights. Selection should then lead to aggressive signals being as harsh and low as possible. It may favour a contrast to this in appeasing and friendly sounds to make them easy to discriminate and so to minimise the chances of their eliciting a hostile response. That signals of opposite meaning are also often opposite in form, presumably to avoid ambiguity, was first noticed by *Darwin* (1872), who referred to it as the principle of antithesis; Morton's idea is similar. Darwin himself pointed to the extreme difference in posture between a hostile dog and the same animal 'in a humble and affectionate frame of mind.'

A similar contrast in seen in gulls, where the beak is shown off in aggressive displays but is hidden, by turning the head away, in appeasing ones. These examples show that signals used to convey similar messages may have features in common. Often this will because the best possible form of signal for a particular function is moulded by

Fig. 6.7. An example of the principle of antithesis used by Darwin. The same dog is portrayed above 'approaching another dog with hostile intentions' and belo 'in humble and affectionate frame of mind'.

environmental constraints but, in some cases, as *Morton* suggests, the form of the signal may be related to the content of the message itself.

Measurement of Communication

Suppose that we wish to learn something about the way crayfish (*Oronectes rusticus*) communicate. If we place crayfish that are strangers together in a tank in the lab, they assume various postures with the large claws and a dominance hierarchy, which is positively related to the size of the animal, develops.

Observation

Our first step is to identify motor patterns that are relatively constant by observing the animals and describing the movements they make. We might divide the behaviour patterns into the following acts: *reteat*—a rapid, backward swimming motion; *cheliped presentation*—the movement of the large claw from a downward-facing position to one that is horizontal to the substrate; *cheliped extension*—the rapid movement of the opened claw toward the other crayfish; and *forward locomotion*—movement toward the crayfish: *fighting*—striking and pinching the other crayfish. Simple observation might suggest that communication is occurring since certain behaviours occur only in the presence of other animals. Perhaps performed by one individual—say, cheliped extension—is usually followed by a particular behaviour performed by the other animal say, retreat. In that case we suspect

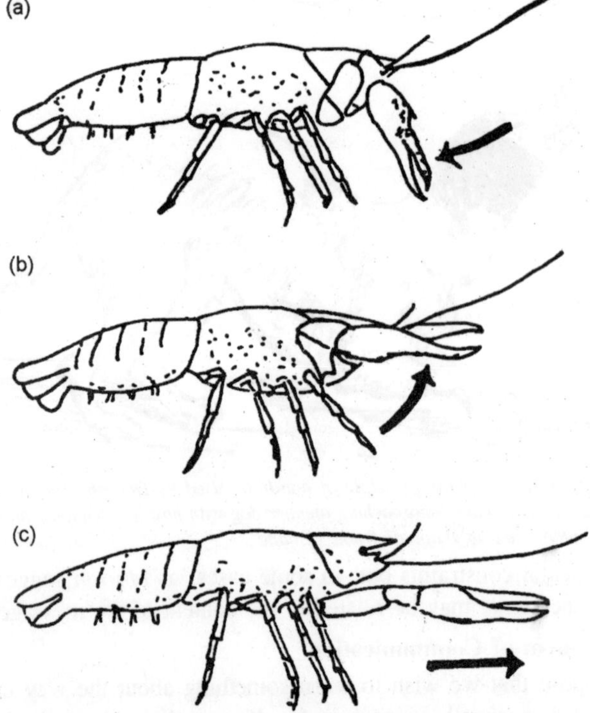

Fig. 6.8. Cheliped (large claw) presentation and extension in the crayfish. In the first phase of cheliped presentation (a), the cheliped tip is lowered and the claw is opened, and in the second phase (b), the cheliped is raised. Part (c) shows cheliped extension. Arrows denote direction of movement.

that communication has occurred. But the occurrence of any behaviour is probabilistic (stochastic). How can we be sure that what one crayfish does affects the behaviour of the other?

Quantification

Observed and *Expected Frequencies*. For a more thorough analysis, we can compare the frequency of each behavioural responses against the expected frequency if the response were random, and without regard to the behaviour of the other crayfish. The results of the chi-square test are presented. Note that some behaviours facilitate response and other inhibit them. Some, such as forward locomotion, seem to have little effect on other behaviours.

Channels of Communication

Odour

From an evolutionary standpoint, the earliest type of communication was chemical, for odour is used throughout the animal kingdom, with the exception of most bird species. Most pheromones are involved in mate identification and attraction, spacing mechanisms, or alarm. The greatest amount of research has been done on insects and mammals. There are several probable reasons why the wide-spread use of chemical signals, has evolved: such signals can transmit information in the dark, can travel around solid objects, can last for hours or days, and are efficient in terms of the cost of production (*Wilson* 1975). However, because these pheromones must diffuse through air or water, they are slow to act and have a long "Fade-out" time. We have a relatively good understanding of insect pheromones. For example, the sex attractant bombykol, produced by the female silk moth, has been isolated, and we are reasonably familiar with the male's perceptual system.

A single molecule of bombykol triggers a nerve impulse in a receptor cell in an antenna of a male. About 200 receptor-cell firings in one second lead to a behavioural response (*Schneider* 1974). The male responds by flying upwind, equalizing the pheromone concentration on both antennae, until he reaches the female. With the commercial synthesis of insect sex attractants, traps baited with pheromone to lure males are used to control such pests as the gypsy moth in the north-eastern United States. Many of the recent studies of pheromones in mammals have been done on rodents. Two general classes of substances, which differ in effect, have been identified: priming pheromones, which produce a generalized response, such as triggering estrogen and progesterone production that leads to estrus; and *signaling pheromones*, which produce and immediate motor response such as the initiation of a mounting sequence. Bronson (1971) suggested that pheromones in mice have the following functions:

Signaling pheromones

Fear substance

Male sex attractant

Female sex attractant

Aggression inducer

Aggression inhibitor.

Priming pheromone

Estrus inducer

Estrus inhibitor

Adrenocortical activator.

In addition, substances produced in male mouse urine speed up maturation in young females *Lombardi* and (*Vandenbergh* 1977). Other urinary products inhibit aggression, increase aggression, stimulate the adrenal cortex, block implantation of embryos, and so on. At this point it is not clear how many different chemicals are involved; different functions may be served by the same pheromone. The sources of these

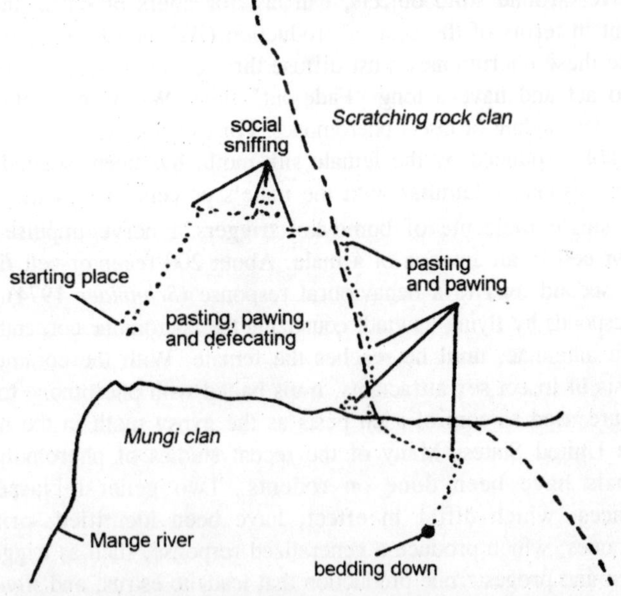

Fig. 6.9. Activities of the Mungi hyena clan along its boundary.

products include the sexual accessory glands and even the plantar tubercles on the mouse's feet. Many of the substances produced by mammals function as a means of staking out territories or home ranges, much as does bird-song.

The advantage of pheromones, as mentioned previously, is that the odour may last for many days ad nights, a significant asset since many animals are nocturnal. Since these substances are often associated with the urinary and digestive systems, eliminative behaviour is often highly specialized. Hyena clans mark and boundaries of their territories by establishing latrine areas. Clan members defecate simultaneously in an area, then paw the ground. The feces turn white and become quite conspicuous. In the same or in another area, hyenas engage in "pasting." Both sexes possess two anal glands that open into the rectum, just inside the anal opening. When pasting, the hyena straddles long stalks of grass; as the stems pass underneath, the animal events it rectum and deposits a strong smelling whitish substance on the grass stems (*Kruuk* 1972).

Sound

Much more information about immediate conditions can be transmitted faster by sound than by chemicals. Sound can be produced by a single organ, and it can also travel around objects, through dense vegetation, and can be used in the dark. Information can be conveyed by both frequency and amplitude modulation. As we can see in sound spectrographs, a unit of birdsong contains sounds of various frequencies and amplitudes. Low-frequency sounds travel great distances and are used by animals with large home ranges. For example, howler monkeys (*Alouatta*) in the Neotropical rain forest signal to other groups with low-frequency calls. Animals with smaller home ranges—for example, squirrel monkeys (*Saimiri sciureus*)—use higher-frequency sounds, which dissipate rapidly. Such calls serve to maintain contact among group members. Aside from human speech, birdsong, whose development has been analyzed be *Thorpe* (1958), *Marler* and *Tamura* (1962), and others, is probably the most complex auditory communication. Birds raised in isolation develop abnormal songs; but if they are exposed to normal song prior to the first breeding season, they modify their songs to match.

Such imitation occurs under natural conditions, and young males develop songs resembling those of their neighbors. Although part of the song develops without any experience, normal, species-specific song must be learned from others, most likely the father. The learning that

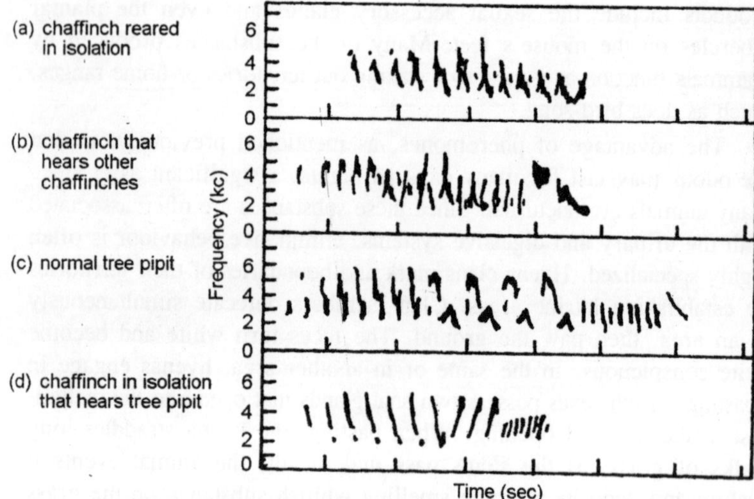

Fig. 6.10. Chaffinch song experiments.

takes place is highly *selective*: hand-reared swamp sparrows failed to learn the song of the closely related song sparrow (both in the genus *Melospiza*) but readily learned the song of other swamp sparrows. Birds may modify their song to match the local dialect if they move into a new area at the start of a breeding season. Females may not mate as readily with strange-sounding males.

High-frequency sounds are used by a variety of animals, particularly mammals. The distress calls of young rodents and some of the vocalizations of dogs and wolves are well above the range of human hearing, as are the echolocation sounds of bats. Although bat sounds are used mainly to locate food objects, communication also occurs

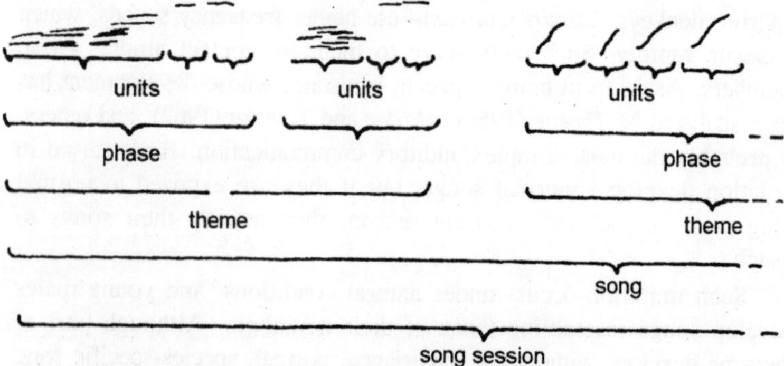

Fig. 6.11. Song of humpback whale.

between predator and prey. Noctuid moths, for example, do not produce sounds themselves, but they possess tympanic membranes on each side of the body that receive sonar pulses from bats. Depending on the location and intensity of sound stimulation, the moth may fly away in the opposite direction, dive, or desynchronize its wingbeat to produce erratic flight. Underwater sound has properties somewhat different from sound in air. Fish and invertebrates produce a variety of sounds, some of which have only recently been investigated.

Marine mammals produce clicks, squeals, and longer, more-complex sounds incorporating many frequencies; most of these are short-duration sounds, which are thought to function in echolocation. Baleen whales (family Mysticeti) produce lower and longer sounds than do the toothed whales, such as dolphins. *Payne* and *McVay* (1971) analyzed the sounds of the humpback whale (*Megaptera novaneangliae*), which are varied and occur in sequences of seven to thirty minutes' duration and are then repeated. The songs have a great deal of individuality, and each whale adheres to its own for many months before developing a new one. Researches have as yet ascribed no clear function to these sounds, but they serve to maintain group cohesion across thousands of miles because of the low attenuation of low-frequency sounds in water.

Touch

Short-range communication in the form of physical contact is used by many invertebrates whose antennae, covered with receptors, are the first part of the body to contact other objects and organisms. Antennae are used by non-social insects, such as cockroaches, and by social insects, such as bees. The honeybee often performs the waggle dance in a dark hive; therefore, much of the information about the type of location of food comes from tactile communication as the worker's antennae contact the dancer and pick up taste cues. Perhaps the most widespread use of tactile displays occurs during copulation. In many rodents stimulation of the back end of an estrous female produces concave arching of the back and immobility (lordosis). In some mammals vaginal stimulation induces ovulation. In most primates grooming is an important social activity and seems to have a function not only in the removal of ectoparasites but also as a "social cement" in the reaffirmation of social bonds.

Most grooming takes place between close relatives, but long-term relationships, indicated by grooming patterns, may exist between non-relatives (*Sade* 1965). In the large Multimale groups characteristic of macaques (*Macaca* spp). And baboons (*Papio* spp.), grooming between the sexes is confined largely to the mating season. South American titi monkeys (*Callicebus*), which live in monogamous groups, entwine their tails when resting. These signals may not to complex, but they are no less important than other signals.

Surface Vibration

Information can be conveyed by patterns of surface vibrations. Males of one species of water strider (*Gerris remigis*) send out ripples of a certain frequency, and receptive females respond by moving toward the source. When a female gets within a certain distance, the male switches to courtship waves. *Wilcox* (1979) also observed that males generate high-frequency (HF) waves when they are close to another water strider. If return HF waves are not picked up from the second strider, the first attempts copulation. In order to demonstrate the function of HF, *Wilcox* used an ingenious playback method to program *females* to send out HF. He glued a magnet to the female's foreleg and allowed her to move freely inside an electrical coil.

When *Wilcox* played an electrical copy of the male HF signal through the coil, the magnet moved the female's leg and she involuntarily sent out the HF signal. When *Wilcox* placed a rubber mask over he male's eyes to eliminate possible visual cues, he found that when the male approached a female, it always attempted to copulate when the female was not sending an HF signal and never did when the female was sending an HF signal. We can apply this type of playback experiment to studies of other types of substrate transmitted signals, such as web communication among spiders.

Electric Field

Some sharks and electric fish have electroreceptors that they use both passively and actively in detecting objects and in communicating socially. Sharks (*Scyliorhinus caniculus*) detect the electric field produced by prey flatfish that are buried in the sand. In addition to electrolocating objects, electric fish of the African family Mormyridae communicate information about species identity (*Hopkins* and Bass 1981), individual identity, and sex by modulating the shape of the electric organ discharge *Moller* (1976) demonstrated that members of this family also use electric organ discharges to maintain group coordination in schools. By altering either wavelength or pulse duration, they can communicate threat, warning, submission, and so on. The advantages of this sensory mode are that it is useful in dark, murky waters, it can travel around and even through certain objects, and it provides precise information on location.

Vision

The need for a direct line of slight and ambient light limit their use, but within social groups, visual displays enable the receiver to locate the signaler precisely in space and time. Monkeys and apes, with a few exceptions, are social, diurnal primates that rely extensively on visual displays. Primates ourselves, we human observers have studies visual systems more than other systems. If you live east of the Rocky Mountains, you have probably seen fields and lawns sparkle with flashes of fireflies. These flashes are emitted by beetles of the family Lampyridae that have specialized photogenic tissue in the abdomen. Such behaviour is known to be related in some way to mate attraction, with each species having its own flash code and flashes of the males varying in intensity, duration, and interval in a species-specific way, as do the females' responses. Within a species the flash interval varies, depending on whether the male is searching for a female or courting one he has found.

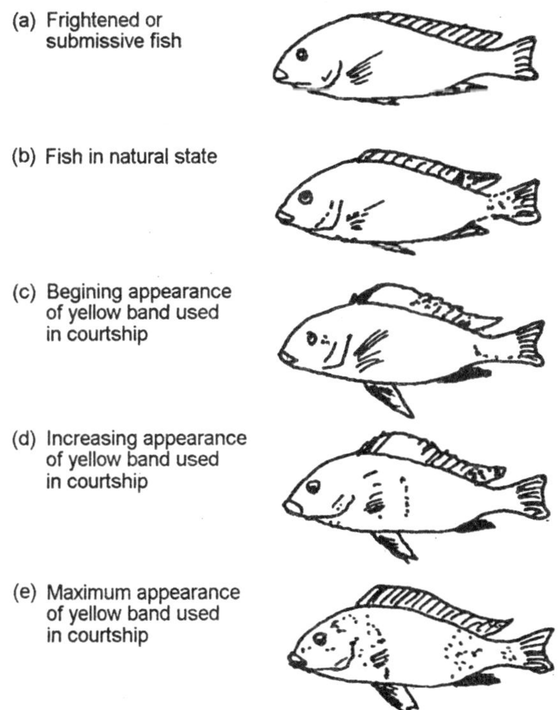

Fig. 6.12. Graded visual signals in the mouthbrooder cichlid fish. An increasing expression of yellow band used in courtship shown in parts (c) through (e).

In one particular species (*Photuris versicolar*), the female, once she has mated, may mimic the flash response of females of closely related species, thereby luring males to her and then devouring them. Such females are aptly termed *femmes fatales* by *Lloyd* (1965). To make matters even more complicated, males of some species of the genus *Photuris* mimic males of other species in order to lure hunting *femmes fatales* of their *own* species into a second mating (*Lloyd* 1980).

In other words, a mated female of species X who is mimicking the female of species Y in order to lure and eat a male of species Y is herself lured into another mating by a male of species X who is mimicking a male of species Y! *Copeland* (1983) argues that more data are needed before male mimicry can be assumed. Fish and some invertebrates are able to change colour within seconds by expanding and contracting chromatophores beneath the skin. The most spectacular in this regard is the octopus; waves of colour advance and recede according to the animal's mood. Because of the limitations of visual displays, these displays are usually coupled with other modes of communication, such as audition. For instance, in the song spread, the redwing spreads its tail, lowers its wings, and raises its epaulets at the same time as it renders the song.

7

HORMONES AND BEHAVIOUR

The analysis of behavioural rhythms is concerned with the same problems of casual explanation whether the period length is measured in seconds, days, or years. The whole gamut of cycles is covered in reproductive behaviour, from the timing of individual acts in a about of sexual behaviour to breeding seasons. Many investigators have attempted to explain the causation of these different levels of organization in reproductive behaviour. In the process they have been confronted with the same difficulty which arises in all studies of motivation, the problem of trying to separate the roles of endogenous and exogenous stimuli. The controversies about factors responsible for the timing of breeding seasons parallel to a remarkable degree those that studies of circadian rhythms have provoked. Here the case will be argued that this parallel is more than coincidental, and that, like circadian rhythms, reproductive behaviour illustrates many of these same principles of behavioural organization.

Precisely coordinated timing is more important in breeding than in any other kind of behaviour. As *Nalbandov* has pointed out. The shedding of eggs by the female frog, for example, must be closely followed by the shedding of sperm by the male. Here the problem is rather simple; all that is needed as a system that will assure the simultaneous shedding of gametes by two individuals of opposite sex (but of the same species) in the same vicinity of the pond at the most propitious season. In more complex animals a whole series of interlocking and synchronized events must follow one another if reproductive efficiency is to be attained. The shedding of gametes, fertilization, gestation, parturition, and lactation are all events that

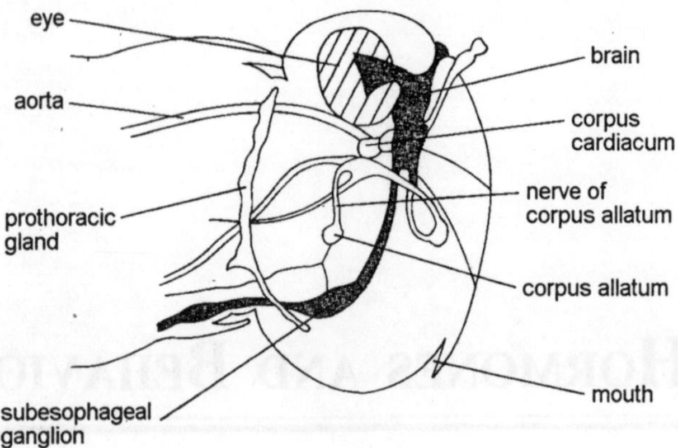

Fig. 7.1. Brain neurosecretory cells and glands of the head region in the grasshopper.

require accurate timing, and all of them must occur in such a way that surviving. Without synchronization and cyclic repeatability, reproduction would become a chaotic and completely inefficient game of chance. Various physiological mechanisms, involving especially the endocrine system, are charged with coordinating this complex series of events.

Reproductive Seasonality

Seasonal Breeding

Breeding season of most species have an annual basis. In the animal kingdom as a whole, reproduction may occur at any time of the year. The periods vary greatly in duration and synchrony. Although

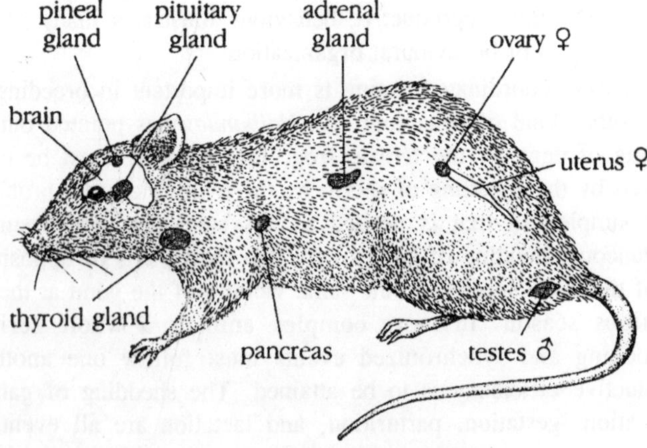

Fig. 7.2. Endocrine glands of the rat.

reproductive capacity often extends over three or four months of the year it is sometimes much more confined. The timing may be extremely restricted, as in the palolo worm that breeds primarily on one or two nights at a particular phase of the moon. Other polychaete annelids breed with similar precision.

The factor dictating the optimal moment for breeding vary widely from species to species to species. Correspondingly, variation may be expected in the mechanism by which timing is achieved. If the critical factors recur in a regular manner, as they don when they are related to the seasons, their predictability permits the evolution of endogenous timing mechanisms. When these factors are essentially unpredictable, as in areas with erratic rainfall, external eliciting stimuli may become extremely important. In most animals the reproductive success is likely to be affected by a variety of factors, some predictable, others not. Most species lie between the two extremes, and depend on both exogenous and endogenous timing mechanisms to set an appropriate time for reproduction. As a further complication, there may be

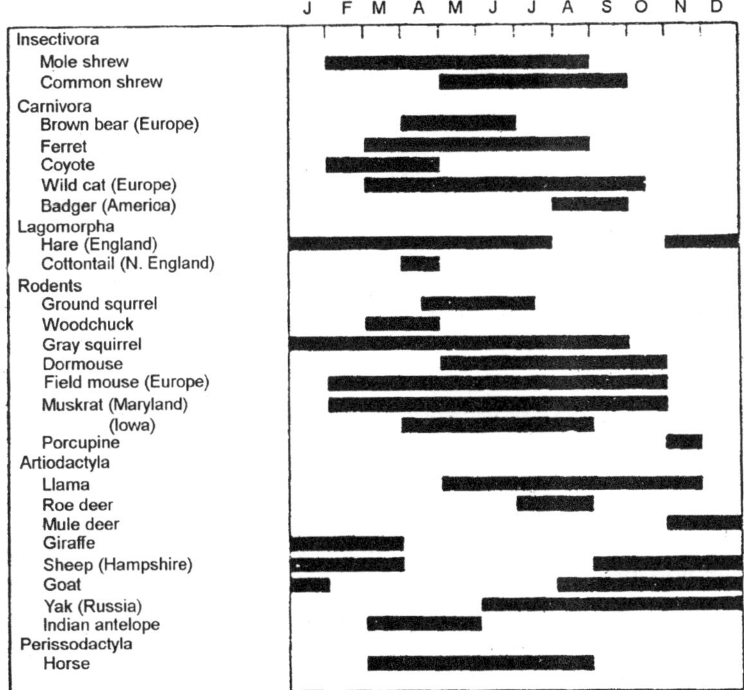

Fig. 7.3. The season of mating activity (black bars) of some representative mammals.

successive mechanisms for coarse and fine timing in the same animal. One set of factors may induce a permissive physiological state of readiness to breed. Another set may regulate the precise time of breeding. A long delay between fertilization and birth poses further problems, as in mammals with long gestation period. An effective environmental trigger for breeding in such species must come at a different season than the optional time for birth of the young.

Exogenous Control

Seasonal breeding is the rule for terrestrial animals in temperate regions where much of the research on animal reproduction is done. It is easy to forget that there are parts of the world where a seasonal pattern is less appropriate. For example, in arid regions many birds and amphibians breed only after unpredictable rainfall. A year may pass with neith r rain for reproduction.

The breeding season of birds inhabiting these areas, such as the red-billed dioch, *Quelea quielea*, of Africa and the zebra finch, *Poephila castanotis*, and other Australian birds, may be quite erratic, lacking any constant relationship to season. When rains are seasonal a more regular cycle becomes established. But here too it need not necessarily be an annual one, as *Miller* (1962) has shown in the Andean sparrow, *Zonotrichia capensis*, which breeds twice a year. A similar situation is believed to occur in some North African birds.

In several species living where rain is meager and sporadic, the gonads of both males and females large for long periods of time, apart from a brief phase of regression after a cycle of breeding. Breeding behaviour appears only when the rains come. Then, within days or weeks depending on the billed diochs begin breeding after rains, in response to the nests.

Older birds seem to be stimulated by a complex of factors, including having an abundance of particular foods to ear, and perhaps even hearing the sound and feeling the contact of falling rain. In these birds which respond to environmental stimuli reproductive behaviour can begin within a remarkably short time. This is true both of red-billed diochs in Africa and of zebra finches which had passed through many months with no rain and no breeding activity at all, courting in the rain of a thunderstorm. Immelmann believes that the falling rain itself directly triggers a reproductive cycle. Some Australian ducks also depend on rains for reproduction, perhaps being stimulated by the level of the flood waters in which they breed.

Evidence of Endogenous Rhythms

A variety of animals go through a cycle of breeding and nonbreeding while kept under conditions which are constant throughout the year, with no cycle of day length. The list includes ferrets, ducks, golden-crowned and white-crowned sparrows, red-billed diochs, budgerigars, zebra finches, and greenfinches. There in no evidence of a precise annual rhythm under such conditions. There is clearly a capacity to cycle, however, which can apparently operate independently of the seasonal change in day length.

The reproduction of ants has been studied over many years by *T.C. Schneirla* Colonies of army ants, *Eciton*, are nomadic and spend much of their time on the move, setting up a new nest each night from which raiding parties go out in different directions in the course of the day. At the day's end there is a mass emigration to another nest site or bivouac. These periods of restless movement, the *nomadic* phase, alternate with a relatively sedentary *statary* phase in which the colony remains in one nest for some days. At this time the queen lays eggs, fertilized by sperm stored since mating at the time the colony was established.

Schneirla has shown that the timing of cycles between nomadic and statary conditions is relatively constant within a species, Egg laying occurs at rather precise intervals throughout the year—every 36 days or so in *Eciton hamatum*. Careful study of periodic changes within the colony shows that the cycle is largely self-generated. The appearance of the new brood excites the colony sufficiently to elicit the next nomadic phase which then continues until the brood has natured. At this time further egg laying by the queen is inhibited, the stimulating effect of the newcomers wanes, and a new statary phase begins. Activity in this phase, together with endogenous changes within the queen, in turn probably contributes to the timing of the next ovulation cycle. The basic period length seems to be the result of interplay between processes and stimuli generated by the organisms themselves, rather than a result of a rhythm imposed by the physical environment.

Lunar Rhythmicities

The Atlantic palolo worm, *Leodice fucata*, the Pacific palolo worm, *Eunice uiridis*, and the fish known as the grunion, *Leuresthes tenuis*, breed annually at certain phases of the moon with almost unbelievably accurate timing. Their behaviour illustrates the interaction of lunar, annual, and daily cycles. Palolo worms live in deep coral crevices, feeding at night by extending part of the body from the hole. Only once or twice each year do they leave the coral rocks to swim freely on the surface. The body breaks in two. The reproductive segment is left in great swarms on the surface, and the other part apparently returns to the coral crevices to regenerate the reproductive parts.

The Polynesian natives gather masses of palolo worms in specially made baskets, eating them raw and collecting great quantities for traditional feasts. Although a few animals may rise on different days and in different months, the bulk of the population swarms on a single day. The natives are reported to predict these occasions far in advance. *Corney's* long run of date on the date of this *great day* affords an insight into the mechanism of this rhythmicity, and allows us to examine the natives' accuracy, and our own for that matter, in predicting a raise in any particular year. The average date of the great rise is November 27, with extremes of November 7 and December 22—an annual cycle of considerable regularity. During the months of November and December there is a sub-cycle, which is in some as yet unknown way timed to the lunar cycle.

Swarming occurs seven, eight, or nine days after the full moon. Between the limiting dates of November 7 and December 22 this particular phase of the moon occurs once in some years and twice in others. The problem is to work out the mechanism of choice in case where it occurs twice, for there seems to be no regular pattern. It may be that some additional exogenous factor is involved. The Atlantic palolo also is most likely to swarm about eight days after the full moon. The time of swarming is in July and falls somewhere within a period of about 30 days in the Caribbean Tortugas. There is again a long run of information, covering 41 years with some gaps, with an intensive study of the actual swarming performances. Environmental factors affect the timing. Wind is particularly important, and when it exceeds 8 miles per hour swarming is inhibited, though sheltered groups may swarm on schedule, out of synchrony with the unprotected population. The time of day when palolos breed is also rather regular. Again *Clark* and *Hess* provide splendid data, demonstrating not only that there is a closely timed schedule within the circadian cycle, but that the schedule is sex dependent, the males preceding the females. Here is a parallel to the social systems of certain vertebrates in which a considerable advantage can sometimes be gained by males by arriving before females at the sexual arena.

A pattern of seasonal reproductive activity that has a lunar rhythmicity also occurs in the grunion. This fish leaves the water to deposit its eggs in the southern California sands. The overall season is long, extending from March through August, but the heaviest runs are in late spring and summer. During a spawning run the fish swim ashore after the high tide has turned, and if females are already there the males remain to mate with them. Milt from one or more males runs through the sand or down the sides of the wet female, the eggs are deposited two to three inches deep in the wet sand, and the next wave returns the fish to the ocean. To be successful, spawning runs must coincide with the highest waters of each tide and with the highest tides of each month. Premature eggs are washed out almost at once. The eggs must develop during two weeks in the wet sand before hatching. Then they are wetted by the next high tide. The timing is usually quite accurate. An individual female may make from four to eight spawning runs on different occasions during the season; so there is a recurring lunar subcycle within the annual cycle. So far nothing is known of the stimuli that time these runs, but the times of spawning are sufficiently predictable to be published in advance for sportsmen.

Day Length

A majority of animals breed on an annual basis, with breeding behaviour suppressed and the gonads regressed for a large part of the year. In most species external timing stimuli play an important role. In non-equatorial species day length is often the paramount factor. Shortening days in the fall may be a cue, both in autumn-breeding animals such as brook trout, *Salvelinus fontinalis*, and in other groups such as ruminants, which begin reproductive behaviour in response to shortening days but to not give birth to the young until a considerably later date. In many spring-breeding fish, mammals, and birds, the combination of long days and short nights in spring triggers the onset of gonad growth. Experimental manipulation of day length simulating spring conditions can readily induce reproductive behaviour out of season.

Interaction between day length and a circadian rhythm may prove to be the critical factor here. *Hamner* finds that house finches show testis recrudescence when a 6-hour light period is coupled with dark periods in a cycle length of 12, 36, and 60 hours but not in cycles of 24, 48 and 72 hours. *Farner* has similar evidence for white-crowned sparrows. Thus the endogenous circadian rhythm may contribute to photoperiodic responses in a very positive way. Another subtle endogenous contribution to regulation of the photoperiodic response comes from the so-called refractory period which follows the breeding

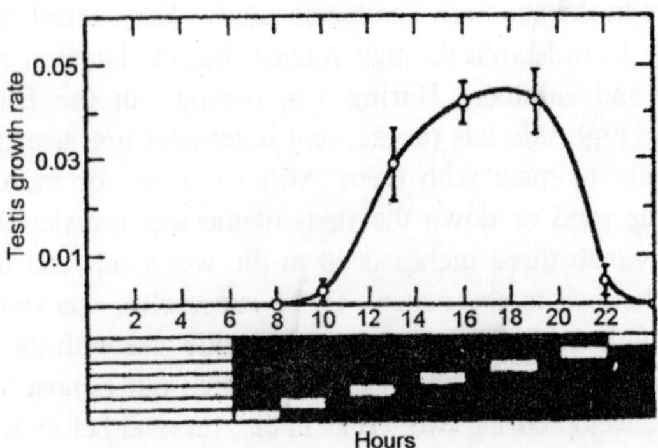

Fig. 7.4. The rate of testicular growth in male white-crowned sparrows kept on 6 hours of light and 18 hours dark, with 2 hours of ligh interrupting the dark period at various chosen times.

season in many birds; this period must be completed before day length changes can elicit another cycle. Its prime function is probably to prevent breeding at an unpropitious time, namely in the late summer and fall.

The refractory period varies in duration from species to species. It may be important in spring-breeding birds which each year migrate across the equator to breed. With a simple photoperiodic response to lengthening days these transequatorial migrants might otherwise be triggered to a breeding response in their winter quarters. The refractory period may extend far enough into the northern spring to prevent this accident. The existence of endogenous rhythms of breeding cannot be doubted. But we should not exaggerate the contribution they make to timing of the breeding season under natural conditions. Susceptibility to the influence of day length is well established. Even the refractory period is not entirely free of effect, for its duration can be varied by changing the pattern of photoperiodic stimulation. In juncos, *Junco hyemalis*, for example, the termination of the refractory period in the fall is hastened by premature exposure to short *winter* days. If instead a bird is left on long *summer* days, the refractory phase may continue into the following spring, so that it will not breed at the proper time. In this species it seems that once the refractory phase is completed a gradual regeneration of the gonads will occur, whatever, the day length regime may be. Exposure to long days, however, will hasten regeneration and must provide the normal trigger for the breeding season in spring.

Thus, there are variations in the effects which a given day-night regime will have at different stages of the endogenous cycle. During the refractory phase, long days will postpone testicular development. A little later, when the refractory phase is passed, the same stimulus accelerates development of the breeding condition. The reverse is true of short days. If we view these phenomena as evidence of interaction between the endogenous cycle and the natural cycle of changing day length, we are immediately reminded of the entrainment of an endogenous circadian rhythm by the cycle of day and night. Again there is a rhythm of changing responsiveness in the course of the endogenous cycle. With circadian rhythms, as with breeding seasons, there is a tendency for some investigators to exaggerate the role of endogenous factors, and for other investigators to minimize it. The solution is surely to be found in a compromise. Both types of influence are required to explain the natural behaviour. Cycles of reproductive behaviour, like circadian rhythms, can be suppressed by certain constant conditions. A less than adequate food supply can inhibit seasonal growth of the gonads of a bird even when all other prerequisites are satisfied.

A steady state, such as the intensity of constant light, has an effect on the period length of a circadian rhythm. We can at least see the possibility that the proportion of light to dark least see the possibility that the proportion of light to dark under a regime of constant day length of the breeding cycle might have a similar effect of extending or reducing the period length of the cycle, and in some cases of eliminating the cycle altogether. Just as with circadian rhythms, experiments on the role of endogenous factors in breeding cycles must take careful account of the constant conditions under which the tests are run, for a rhythm may appear in one circumstance and be inhibited in another. Many aspects of breeding seasons and their control have been omitted from this review.

Weather plays a role in accelerating or retarding the spring breeding seasons of many birds which make their coarse adjustment to the season from photoperiodic cues. When it is fine and warm the breeding of birds begins earlier in the spring, and occasionally may even occur in the fall as well. Temperature is a critical factor in some mammals such as ground squirrels and, of course, in poikilotherms. An increased food supply may trigger breeding in some cases (e.g., crossbills, *Loxia curvirostra*, and there are many more complications to which attention has been drawn. The basic principles presented here serve to reinforce our thesis. Erratic and unpredictable breeding seasons are triggered by external stimuli, with endogenous factors usually playing no more than a permissive role in the timing mechanism. In regular annual breeding seasons, endogenous factors assume a more prominent part. The effects are achieved in nature not so much by endogenous triggering of breeding as by endogenous modification of the response to day length and other external stimuli, at different phases of the cycle. In such cases the cycle cannot be regarded as either endogenous or exogenous but must be seen as the result of a combination of both. Thus, it makes little sense to argue one side to the exclusion of the other.

Mating and Swarming

To comprehend the short-term patterning of events which will be discussed later, some brief comments are needed on the social framework within which these events occur. Social organization has a special bearing on the matter of timing, since the overall social system determines to a large extent the amount of latitude that is permitted in sexual activities.

Mating Swarms

The greatest premium on precise timing occurs when masses of individuals gather together with promiscuous shedding of gametes. For examples, in the palolo worm, discussed earlier, mating is restricted

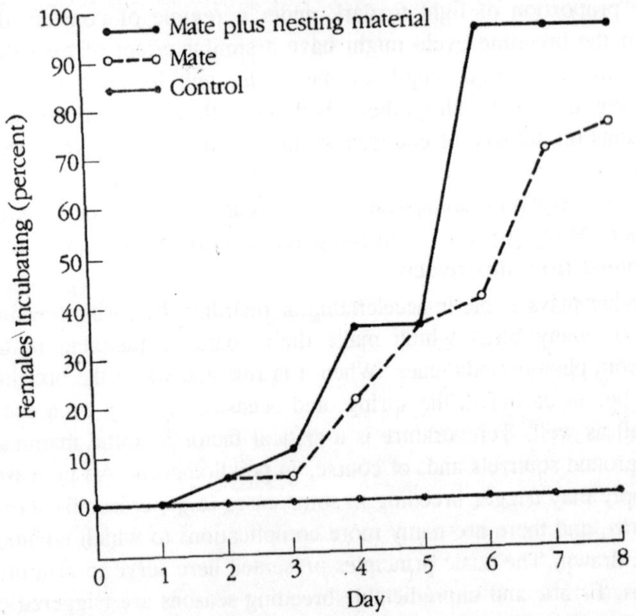

Fig. 7.5. Development of incubation behaviour in female ring doves is effected by associatio with a mate or with a mate plus nesting material.

to one or two days of the year, and the mating swarms form in the predawn hours. Pelagic fish may form similar promiscuous swarms, with mating synchronized to a particular part of the day. The sardine, *Sardina pirchondus*, for example, spawns in the early evening hours. Pelagic invertebrates solve the problem by emitting vast numbers of male gametes almost continuously during the breeding season. The environment is thus saturated with male gametes, and the chances of released eggs being fertilized are high. The effectiveness of external fertilization depends largely on spatial distribution of the mating animals. The distribution is accomplished in numerous ways. In some marine organisms the male and female are attached to each other. Other mechanisms to ensure temporal synchronization of mating rely on either synchronizing environment stimuli or direct communication between the sexes. The find adjustment in timing is often made by chemical signals or 'pheromones', emitted by ripe females, which cause the release of spermatozoa and guide them to the ovum of the female.

Mating Locations and Pair Bonds

The males of some fish species establish locations where gravid females visit them for spawning then the females are driven away. Elaborate ritual can play a key role in such exchanges, and signals can be transmitted through several sensory modes. Increased development of such ritual seems to be correlated with the use of smaller numbers of larger eggs. Nevertheless, there is seldom any extended mutual interaction between the sexes in most fish such as is common in birds and mammals. Some degree of promiscuity seems to be the general rule.

Internal fertilization in vertebrate and invertebrate groups often involves a larger investment in particular acts of fertilization and a corresponding economy in the output of gametes. Frequently it demands a closer relationship between individual males and females. If the male drops a spermatophore which is picked up later by the female, as in some insects, there is less demand for precise sexual coordination than when gametes are transferred directly. Internal fertilization is most characteristic of terrestrial animals in which copulation transfers sperm from males to females.

Other behaviour is associated with embryo retention and maximal investment in the success of fewer young. Pair bounds may extend through a breeding cycle, a season, or sometimes through life. Males frequently protect the pregnant or incubating female, and in some species the involvement of both sexes in postnatal care of the young results in complex behavioural interactions among members of the family. There are many exceptions to these generalizations.

In some birds, certain grouse for example, the pairing bond between male and female is as brief as that in many fish. In other birds the contact may last for several days during the laying of the eggs. Often it lasts for several weeks or months, and sometimes for life. Similarly, among mammals there is a complete spectrum, from the brief contact between male and female and an exclusion of the male from the family in hamsters, to the persistent family group in beavers. But amphibians and fishes, even viviparous species with internal fertilization, rarely show the extended and elaborate sexual contacts that occur in mammals and birds.

Delayed Implantation

Sheltering the embryo within the mother's body during early development involves a commitment to a long, intricate, and rather rigid series of physiological processes, which complicates the problem of timing the breeding season. Some mammals have evaded the problem by delaying the implantation of the embryo in the wall of the uterus for various periods of time after fertilization. In the fur seal, *Callorhinus*, for example, mating takes place immediately after the young are born on the summer breeding islands. Only eight months are needed for embryo growth.

Development does not begin until implantation occurs, long after the events of ovulation and fertilization. Delayed implantation postpones the timing of parts until the return to land in the following year when copulation also occurs. In other species, such as some mice delayed implantation may delay the physiological burden of pregnancy while the female is still lactating. Yet mating occurs in the first two days after the birth of the young, as in mice which lack delayed implantation. Delayed implantation thus provides a mechanism whereby the timing of mating can be emancipated from the energy requirements of pregnancy.

Exogenous factors can intrude in determining the timing of implantation. In the mink, *Mustela vison*, and the marten, *Martes americana*, *Pearson* and *Enders* have shown that the timing of the initiation of development of the unimplanted embryo is subject to photoperiodic manipulation. In the marten, for example, mating occurs in July and August. By artificially lengthening the days from September onward the time of birth can be advanced from April to the previous December, cutting short the implantation delay.

Short-term Reproductive Cycles

Patterns of Ovulation

The temporal organization of ovulation in female vertebrates follows a number of different patterns. In amphibians and fish the female generally comes into seasonal reproductive condition, and then she ovulates and engages in sexual activity with the male for a limited part of the reproductive season. Her sexual receptivity rises in response to stimuli from a variety of sources, both environmental and social. In mammals ovulation and sexual receptivity recur periodically as a manifestation of the estrous cycle. The cyclic receptivity of certain fish suggests the sort of step that may have anticipated this development. Embryo retention within the uterus may have been a primary factor favouring the development of this cyclicity since successful implantation requires a suitable condition of the uterine wall, and this has to be prepared before hand.

Physiological Basis of Ovulation

The physiological mechanisms underlying mammalian estrous cycles involve both endogenous and exogenous factors. There is a basic feedback relationship between estrogen from the ovaries and the *Follicle Stimulating Hormone* (FSH) from the anterior pituitary. Negative feedback results in a cycle of ovarian activity and sexual behaviour in mammals which ovulate-spontaneously. There is also evidence of a separate cycle of hypothalamo-hypophysial activity, for cycling of gonadotropin production has been recorded in female rats and mice after destruction or removal of all ovarian follicles.

Some mammals have one period of heat per year. An example is the silver fox, *Vulpes fulva*. The females are receptive only during a period of from one to six days during February. With some exceptions, such as the basenji with one annual heat, most domestic dogs have two periods of heat per year, one in the early spring and one in the fall. A bitch is then receptive for one or two weeks. Many mammals are polyestrous, having a series of receptive periods. These periods follow one another at regular intervals throughout the breeding season, varying in duration from a day or less in rodents to a week or more in primates.

The estrous cycle of the female rat lasts four to five days. It is associated not only with variations in sexual receptivity but with striking changes in overall activity which are readily detectable if the rate is placed in a running wheel. In most mammals the period of maximal

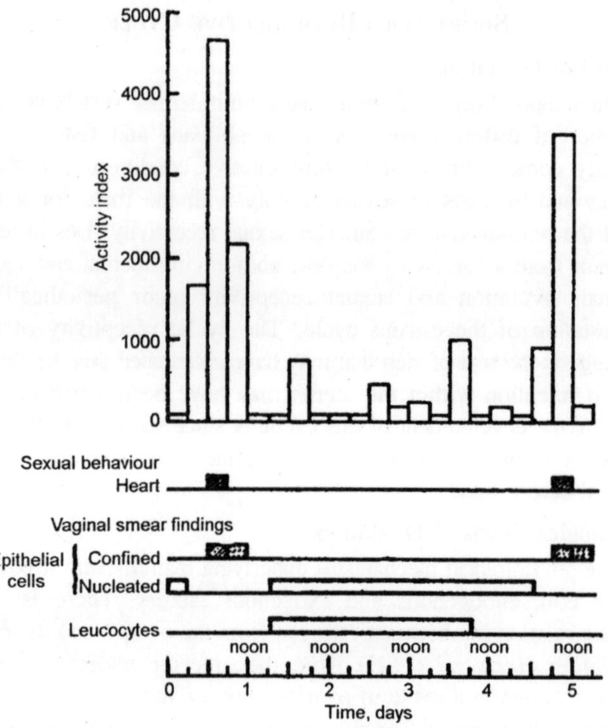

Fig. 7.6. The activity of a female rat can be correlated with changes in reproductive physiology.

receptivity seems to just precede the time when fertilization is most likely to occur, some hours before ovulation. For example, female golden hamsters, *Mesocricetus auratus*, kept on a 12-hour day-night schedule came into heat between 1 hour before the onset of darkness and about 5 hours after, and ovulated 8 or 9 hours later. A precise temporal correlation between maximal receptivity and ovulation is not universal, however, as *Rowell* has shown with rhesus monkeys, *Macaca mulatta*. Female rats and hamsters, both 'spontaneous ovulators,' will shed the eggs whether they have copulated or not. As we shall see, there are mammals, called *induced ovulators*, in which ovulation is triggered by external stimulation.

Ovulation in Birds

Female birds differ from mammals in lacking any strict parallel to the estrous cycle. As already pointed out, the general season for breeding is dictated by one set of factors, both internal and external, and the precise timings is regulated by other factors. In birds the ultimate control of this detailed timing is largely the prerogative of the female. Such factors as weather, availability of nesting sites and food, and social stimuli from the male play a significant role in determining the actual day when a female bird will ovulate. Once a clutch of eggs is started, laying proceeds at a regular interval until the clutch is complete. The raising of a brood may be followed by a further cycle of ovulation. Another clutch may also be produced if a nest is destroyed.

Most of the evidence on the physiological basis of ovulation in birds is concerned with chickens, although *Hinde*, *Lehrman*, and others are extending our knowledge of other birds as well. Some intrinsic mechanism ensures that follicles mature in a graded series, even when follicular development is evoked by massive gonadotropin injections. The follicle-stimulating hormone, FSH, provides the stimulus for this maturation. We may assume its production through a large part of the breeding season. Early in the season the growth of ovarian follicles is slow, but some days before laying occurs there is rapid acceleration, presumably in response to pituitary gonadotropins. External stimuli of various kinds control of timing of the events that lead to the lying of the first egg, special significance being attached to those stimuli generated in courtship and nest building activity.

Courtship and Ovulation in Birds

A number of social factors can contribute to the occurrence of ovulation in a female bird. The best evidence is from pigeons. An isolated female pigeon normally will not lay eggs, but she can be induced to do so by allowing her to see another pigeon, or even by giving her a mirror. The relationship between external stimuli and the breeding cycle has been pursued further in a long and ingenious series of experiments by *Lehrman* and his colleagues. They have shown that participation of ring doves in courtship hastens the onset of nest building.

Apparently this behaviour evokes the release of estrogens from the ovarian follicles, an effect that must be mediated by the pituitary. Stimuli generated by male courtship are important for gonadal development of female doves. Caged females with a view of castrated, noncourting males show signs of less estrogen production than caged females with a view of normal, courting males. The next phase of the cycle, involving nest building, is induced by the secretion of estrogens. The action of nest building and the stimuli that they generate serve in turn to hasten ovulation, apparently by eliciting the release of Luteinizing Hormone (LH) from the pituitary; increasing production of

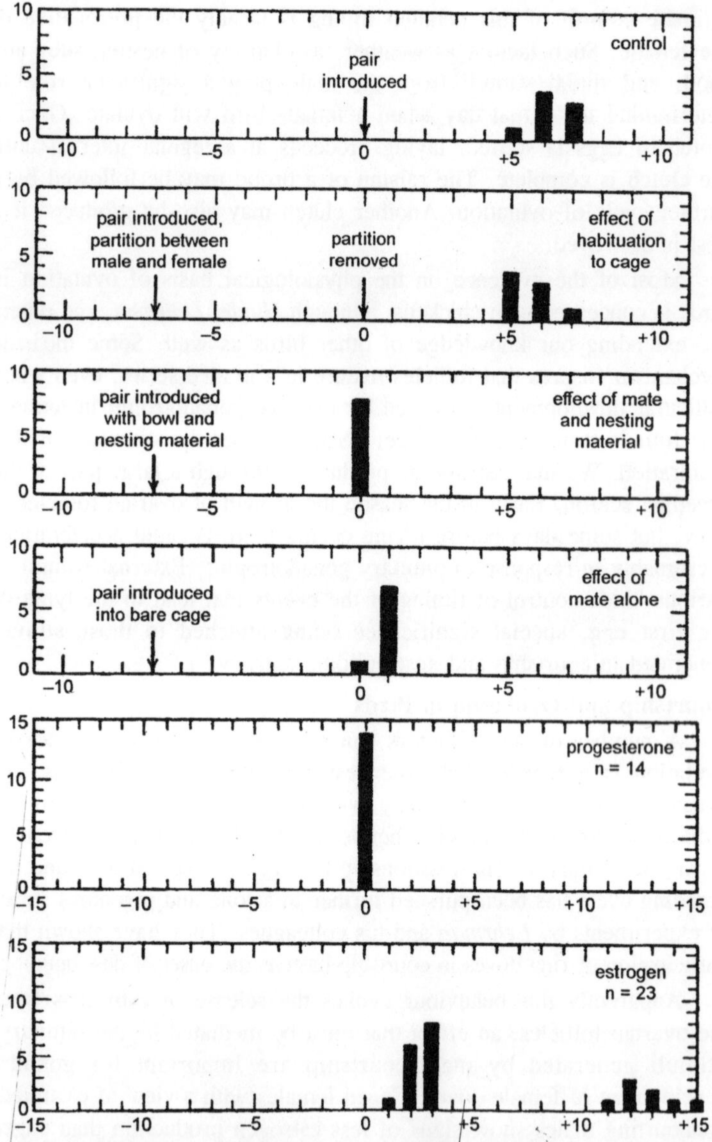

Fig. 7.7. Incubation latency of ring doves with various environmental and hormonal pretreatments. A nest with eggs was introduced on day 0.

progesterone by the ovaries may also occur. The stage is then set for ovulation. According to Lehrman, progesterone facilitates another burst of LH production, and the first egg is laid.

The rhythm of further egg laying until clutch completion varies in a characteristic manner from species to species. Evidently the presence of an egg in the oviduct inhabits the production of LH. As the egg is laid the release of LH follows rapidly (with a few minutes in the domestic hen), and the sequence of events leading to ovulation and laying of the next egg is initiated.

The cycle of laying is also influenced by a daily rhythm. Birds and to lay eggs at a particular time of day which changes with a shift in the rhythm of light and dark.

Incubation Patch

Shortly before the start of incubation, many female birds develop one or more; incubation patches' on the breast and belly. Down feathers are shed, the epidermis thickens, and there is a great increase in vascularity and edema in the underlying tissues. Increased production of estrogen, acting with luteotropin (LTH) from the anterior pituitary, and perhaps with progesterone, seems to cause these changes. Experimenters using hormone therapy have not yet been able to develop incubation patches comparable to those occurring in nature. *Selander* and *Yang* suggest that full development may depend either on a special sequence of hormonal actions or on tactile stimulation of the patch, such as might be provided by the eggs during incubation.

In phalaropes some role of male and female are reversed, and male incubate. In two species, *Steganopus tricolor* and *Lobipes lobatus*, *Johns* and *Pfeiffer* found that testosterone, rather than estrogen, is the

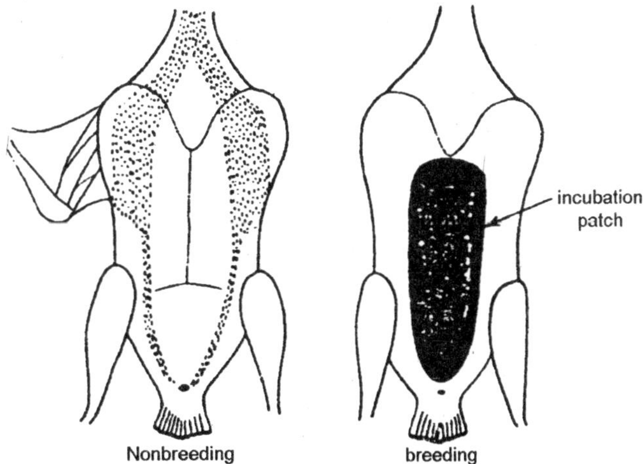

Fig. 7.8. Female white-crowned sparrow, showing the position of the patch.

effective gonadal hormone in incubation patch development and that it acts together with LTH. Although females do not normally develop an incubation patch, they do so if treated with testosterone and LTH. In a few passerine species both males and females have incubation patches, but the hormonal mechanism remains to be worked out.

Clutch Size

The number of eggs in a clutch, a species characteristic seems to be controlled by two different mechanisms. In some species removal of eggs from the nest as they are laid results in the continued production of eggs. These species are called 'indeterminate layers.' Perhaps stimuli generated by the nest and eggs, received in part by the sensitive incubation patch, play a role in limiting the size of the clutch. Development of the incubation patch is accompanied by a measurable increase in tactile sensitivity. Visual stimuli from the eggs may also play some part. With 'determinate layers,' on the other hand, removal of the eggs does not increase the number of eggs laid, and endogenous factors are more prominent in determining clutch size.

Like an indeterminate layer, however, such a bird will produce fewer eggs if some are introduced into the nest prematurely by the experimenter, and she may even refrain from laying altogether. Thus determinate layers are responsive to stimuli from the nest and from the eggs as well. In some species the situation is complicated by the commencement of incubation at the start of laying, not toward the end as is more usual. Laying seems to terminate after a certain number of days of incubation. This review of events leading up to completion of the clutch serves to emphasize the intricate experimentation that is required to disentangle the roles of internal and external factors in the natural sequence of behaviour. The same is true of later stages of the cycle.

Progesterone seems to induce the start of incubation. As soon as incubation begins, the third pituitary gonadotropin, Luteotropic Hormone or prolactin (LTH), is released, facilitating further incubation. LTH also inhibits the production of FSH and LH and therefore inhibits the production of the gonadal hormones as well. Then the sexual phase gives way to the parental phase. LTH, acting with ovarian steroids, furthers development of the brood patch and also encourages care of the young after they hatch, particularly brooding behaviour, feeding, and defence from predators.

Doves are specialized to produce '*crop milk*' at this time, as a result of proliferation of the crop wall. The milk is regurgitated by members of both sexes directly into the mouth of the young. The production of crop milk is dependent on LTH and constitutes a standard bioassay for this hormone. Although LTH provides the normal basis for parental behaviour, some parental behaviour can be evoked in its absence. Presenting adults with young will elicit parental behaviour in some animals but does not produce crop milk in pigeons. In time, the young become independent and the level of LTH production declines. If the season permits, another cycle of FSH production and gonad growth may begin and a further breeding cycle will ensure. If not, the pituitary will not produce more gonadotropins until internal and external factors once more combine to instigate the next annual breeding cycle.

Precopulatory Behaviour

There are certain conditions that precopulatory behaviour must satisfy. The need for synchronization of the reproductive physiology of male and female has already been mentioned. Before this is possible, the two sexes must find each other. If there is to be effective reproduction, each adult animal must locate a partner of the opposite sex and of the same species. In order to avoid wasting time and gametes there must be reproductive isolation.

In many animals the problem of reproductive isolation is settled well before the two sexes engage in reproductive behaviour and copulation. A particular temporal or spatial distribution of activity may ensure minimal contact with other species. Mates may be selected from a pre-existing social group; this is often the case with mammals in nature. The processes by which the group become established at an earlier time may already have excluded other species. In many monogamous birds the pair bond is established some weeks before mating.

On the other hand, there are animals from all levels of the animal kingdom in which the sequence of behaviour leading to copulation included within its functions that of ensuring reproductive isolation. When the sexes have only a brief contact, we find the most extreme developments in precopulatory behaviour of patterns of sound, odour, appearance, and physical contact. The requirement for strong species distinctiveness seems to be at lease one of the functions of the elaborate nature of precopulatory behaviour. Some insects may solve the problem at a later stage in a direct mechanical fashion, by the evolution of specific 'lock and key' types of genital apparatus, though the evidence for this equivocal. But in most terrestrial organisms the mechanisms for reproductive isolation involve a complex of behavioural traits that are effective before actual mating, such as selection of habitat, timing of activity, and choice of a mate.

With some exceptions most of the information about the structure of reproduction behaviour in small mammals is necessarily derived from experiments in which two animals are placed in close contact in a small cage. This procedure bypasses whatever forms of distance communication may be involved in the establishment of contact, as well as environmental barriers in nature that might hinder it. We shall return to this subject in discussing external stimuli and reproductive behaviour.

Copulation

The problem of ensuring physiological synchrony between mating partners varies widely in different animals. Males of many species are ready to copulate at short notice during a long season. They must induce readiness in the female by various kinds of precopulatory courtship activity. The need may also be mutual. The contrast is well illustrated by comparing the precopulatory behaviour of bulls and horses.

The stallion differs from the bull in having a typical vascular penis in marked contrast to the fibro-elastic penis of (the bull). In the non-erect condition the horse's penis is quite flaccid and is withdrawn within the confines of the prepuce. The retractor penis muscle is relatively undeveloped and adherent to the ventral surface of the penis. There is no sigmoid flexure. Erection and protrusion of the penis are affected by gradually increasing tumescence of the erectile vascular tissue in the corpus cavernosum penis. Erection usually takes place rather slowly, and depends upon the continued reception of erotic stimuli derived from courtship and foreplay. Foreplay appears to be an essential accompaniment of vascular erection and this is well featured in the behaviour pattern of the stallion.

By contrast with the stallion the bull exhibits little or no foreplay prior to mounting and copulation. The penis of the bull is of the fibro-elastic type. It is of small diameter and relatively rigid even in the non-erect condition. The erectile tissue is small and the penis undergoes little enlargement on erection although it becomes more rigid. Protrusion is affect partly by erection but mostly by relaxation of the retractor muscle and straightening of the sigmoid flexure. It mating with the cow there is little or no foreplay. The bull tests the cow by smelling or exploring the vulva with muzzle or tongue and by placing the chin on the (base of the) cow's tail... If the cow is not (in) heat, she moves away rapidly and evades the bull's attempt to mount, if she is (in) heat she stands still and accepts service, which is performed with great rapidity.

In Polyestrous mammals, which provide most of our data on copulatory behaviour, precopulatory exchanges are relatively brief. They seem to function primarily to allow the female to reveal whether or not she is receptive and to provide sufficient mutual stimulation to bring both parties to the appropriate level of motivation for copulation to occur. After preludes are completed copulation takes place. The actual temporal sequence of activities varies from species to species. During sustained association between a pair of animals a number of copulations is likely to occur during the estrous phase of the female, maximizing the possibility that fertilizations will take place.

Induced and Spontaneous Ovulation

In certain mammals, copulatory activity is necessary for ovulation. Such animals are referred to as induced ovulators rather than spontaneous ovulators. The domestic rabbit and cat are typical induced ovulators in which the ovarian follicles degenerate at the end of each estrous cycle if copulation does not occur. In the short-tailed shrew, *Blarina brevicauda*, several copulations are required before ovulation will occur. In the mink, *Mustela vision*, participation in precopulatory behaviour may suffice. In both species the courtship is rather violent affair.

Immediately following copulation, some birds perform strikingly elaborate displays. In ducks these postcopulatory displays take the form of rapid swimming of the male, with head held high. In the avocet the mating birds cross their bills and race side by side for several yards. No adequate explanation for these displays has been put forth, though it seems possible that elaborate and precise patterns such as these have an important function, possibly in triggering ovulation. Other external stimuli can play a part. The characteristic beginning of estrus at a particular time of day has already been pointed out. Experiments with shifted day-night rhythms have shown that light-dark stimuli are important in establishing the timing of ovulation. A change in the day-night regime is followed by a gradual phase shift of ovulation to harmonize with the new circumstances. The importance of external stimuli to induce ovulators is obvious, although the female must be in the appropriate physiological state for ovulation to occur. In some mammals, the spontaneous ovulations, endogenous factors are dominant, and in others, the induced ovulators, the effects of external stimuli are more prominent.

Sequences of Mating Behaviour

Having considered the organization of reproductive behaviour on an annual basis, as well as the shorter cycles which are to be found within the breeding seasons, we can now turn to the temporal organization at the bout level. What are the preludes and sequels to the act of insemination in different types of animals? How is the whole sequence organized in time. And what is known of the mechanisms by which this temporal patterning is achieved? There is so much variation within the animal kingdom that it almost defies analysis, but there are some functional considerations, relating to the appetitive and consummatory properties of sexual behaviour, which help to introduce order into the wide range of possibilities.

In rats and guinea pigs copulatory activity is organized in bouts, in much the same way as other types of behaviour. In general, sequences of appetitive precopulatory behaviour lead to the consummatory activity of copulation. This is followed by a period with sexual activity absent, perhaps with the intervention of other activities, before another bout of sexual behaviour begins. Complications within this simple framework have occupied the researches of many investigators, notable *F.A. Beach* and *W.C. Young*, over a number of years. In the guinea pig the sequence is fairly straightforward.

When a male encounters a female in heat, he follows her, sniffing at her anogenital region. The female is likely to adopt the lordosis posture if she is receptive; then he mounts her, and perhaps after some abortive attempts, achieves intromission and makes pelvic thrusts. He may dismount and repeat this sequence several times before he achieves ejaculation, when he fails back in a distinctive manner. Both animals clean their genitals and further sexual activity is then suspended for a variable length of time. In rats and hamsters the preliminaries to copulation are similar to those of the guinea pig. Grooming behaviour is often included, an activity which is prevalent in the sexual activities of many rodents. The later stages are more complex. Mounting and pelvic thrusts are normally repeated several times before the male achieves ejaculation. Then the sequence comes to an end and is followed by 24 or more hours without sexual activity.

Careful quantitative description of the temporal sequence of events shows that the concepts of appetitive and consummatory behaviour are applied less easily to rats and hamsters than to the guinea pig. Successive ejaculations are indeed followed by increasingly long *postejaculatory refractory periods* as the bout proceeds. However, we must also consider the paradoxical finding that proceeds. However, we must also consider the paradoxical finding that successive ejaculations are achieved with fewer and fewer intromissions within the same series. There are in fact suggestions of increasing and decreasing motivation at the same time, reminding us once more of the dangers of considering motivation in unitary terms. Evidently a sequence of intromissions and ejaculations has both inhibiting and facilitating effects on subsequent sexual activities.

It is only when the inhibitory influence becomes completely dominant that the bout is terminated. Artificial separation of the pair between the first and second ejaculations for varying periods of time shows that the facilitating effects of the first ejaculation wane rapidly, returning to the original level after about 90 minutes. As the inhibitory effects of successive ejaculations increase, the refractory period presumably exceeds this period of facilitation, and the sequence terminates. It is only when the inhibitory influence becomes completely dominant that the bout is terminated. Artificial separation of the pair between the first and second ejaculations for varying periods of time shows that the facilitating effect of the first ejaculation wane rapidly, returning to the original level after about 90 minutes. As the inhibitory effects of successive ejaculations increase, the refractory period presumably exceeds this period of facilitation, and the sequence terminates.

Beach proposes at lest two distinct physiological mechanisms to explain the temporal patterning of mammalian sexual behaviour, and other types of evidence point to a similar conclusion. Beach postulates an 'arousal mechanism' which mediates the initiation and maintenance of sexual excitement and underlies the appetitive phase of the sequence. The execution of copulatory and ejaculatory reactions is held to depend on the 'copulatory mechanism' called into play when activity in the arousal mechanism becomes sufficiently intense. No doubt a multiplicity of physiological effects will prove to be involved.

Beach believes that the copulatory mechanism in rodents is particularly involved with spinal and autonomic elements and the hypothalamous, and that the arousal mechanism involves the cerebral cortex, especially in males. The latter point is suggested by the ease with which learning can affect the arousal responses of male rodents to external stimuli. In higher mammals cortical involvement becomes still more extensive. As we shall see later, the sex hormones may even be dispensable in experienced males.

Mechanisms of Terminating Sequences of Reproductive Behaviour

In sexual activities, as in other types of behaviour, there is need for a combination of facilitating effects to sustain bouts of activity and inhibitory effects to bring sequences to an end. The mechanisms available are as varied as, for example, those that regulate feeding behaviour. Intrinsic consequences of sexual activity, both facilitating and inhibitory, seem to contribute to the control of behavioural sequences in rats. Other factors play a part as well.

The account thus far has been of a sequence of responses of a male rat or hamster to the same estrous female. In fact, many such experiments involve the repeated substitution of one estrous female for another. These substitutions provide the maximum intensity of external stimulation for the male, necessary because males become habituated to repeated contacts with the same female. The effect was clearly demonstrated in guinea pigs and rats by allowing a male to mate with one estrous female and then reviving sexual activity to some extent by substituting a new female. Habituation makes a definite contribution to termination of a bout of sexual activity involving the same two animals.

In rats and guinea pigs the effects evidently act in concert with intrinsic consequences of copulation. In some other animals habituation seems to take on a still more dominant role. For example, experiments designed to obtain large quantities of semen from bulls for artificial insemination have shown that bulls too wane rapidly in their responsiveness to the same stimulus object—a cow or a dummy. If on the other hand, the stimulus situation is constantly changed by using a new animal as the dummy, or by presenting the old one in a new situation, a bull will continue to perform a remarkable number of consecutive ejaculations. "Sexual satiation to one stimulus provides negligible interference with a subsequent response to a new stimulus animal." Evidently intrinsic consequences of motor sexual activity contribute little to the normal termination of a sequence in this case.

Barrass has described a somewhat similar phenomenon in an insect, the chaclid hymenopteran *Mormionella vitripennis*. A female normally ceases to be receptive after a male has successfully copulated with her. So a particular bout of sexual behaviour is naturally terminated in this way. The provision of new receptive females revives the male's sexual activity, however, and *Barrass* concludes that "mating in this insect does not result in reduced ability to repeat the same behaviour immediately." New nonreceptive females will also elicit some revival of male courtship, but the effect gradually wanes with a series of presentations. In this species habituation cooperates with postcopulatory changes in the behaviour of the female to terminate a bout of sexual activity between a given pair of animals.

Changes in the external stimulus situation generated by sexual activity contribute to the termination of bouts of sexual behaviour in other species. Some of the best-established cases occur in fishes. Eggs laid by the female at the climax of a courtship sequence provide stimuli which inhibit any further sexual activity by the male; he is now aggressive toward the female. In the bittering, *Rhodeus amarus*, the male defends a territory around a mussel and is visited by the female, who lays the eggs in the mantle cavity, *Wiepkema* has shown that it is not the act of sperm ejection which terminates a bout of male courtship. Rather it is the presence of eggs in the host mussel. In the three-spined stickleback, *Gasterosteus aculeatus*, deposition of eggs in the male's nest inhibits further male courtship of the female. The same change in the external stimulus situation also facilitates other patterns of behaviour, including sperm ejection over the eggs and attack upon the female. This example is a reminder that the same sequence can be appetitive with respect to some behaviour patterns and consummatory with respect to others.

We have already reviewed Lehrman's demonstration of the way stimuli generated at one phase of behaviour can lead to the next phase in the reproductive behaviour of ring doves. Another example is provided by Hinde's analysis of nest building in canarics. Nest building is a complex task involving the selection of several types of material. This selection shifts from grass to feathers as the day of the first egg approaches, and is paralleled by a change in the frequency of different nest-building movements. *Hinde* has shown that the progressive changes can be explained partly in terms of endogenous hormonal changes and partly as a result of stimuli received from the decrease in the size of the nest cup during construction. The texture of the nest material also plays a part, and so apparently does the increasing tactile sensitivity of the bird's incubation patch.

The interactions between changing exogenous and indogenous factors are so complex that a simple application of the ideas of appetitive and consummatory behaviour is not often easy. Nevertheless, there is a widespread general pattern to bouts of sexual behaviour, and there are various means of ending or consummating a sequence. Many different mechanisms can contribute to the emergence of similar system of temporal organization in the overt behaviour.

Table 7.1. Mechanisms for termination of a bout of male copulatory activity with a given female

Animal	*Habituation of male to individual female*	*Some change in female after mating*	*Change in environment after mating*	*Consequence of performing sexual motor activities*
Rat and guinea pig	+	–	–	+
Bull	+	–	–	–
Stickleback	–	–	+ (eggs in nest)	–
Mormionella	+	+	–	–

MATERNAL BEHAVIOUR IN MAMMALS

Pregnancy

In the absence of fertilization, estrous cycling in polyestrous female mammals continues. If mating occurs, the course of the cycle is broken. The series of changes which follows leads to implantation of the embryo in the uterine wall and the provision for nutrition, birth, and suckling of the young together with all the associated behaviour patterns of nest building, defence, and care of the young the timing of these events varies widely from species to species. The break in the estrous cycle results from a change in the relationships between the anterior pituitary and the corpus luteum, a structure derived from the ruptured follicle left behind after ovulation.

In some mammals there is no functional (i.e., progesterone-secreting) luteal phase in a normal estrous cycle. In others the corpus luteum persists for some days. Under the influence of LTH it produces progesterone which, together with estrogen, prepares the uterine wall for the implantation of embryos. Eventually the corpus luteum ceases to secrete, and cycling begins again. The effect of fertilization is to instigate or prolong the development of the corpus luteum. It may persist either until the end of the gestation period or until the placenta is producing sufficient quantities of estrogen and progesterone to take over the functions of maintaining the condition of the uterus and inhibiting further estrous cycles. Normally, persistence of the corpus luteum results from arrival of the embryo in the uterus. The embryo causes a persistence of LTH production by the anterior pituitary which then influences the corpus luteum. The further production the LTH also combines with estrogen and progesterone in encouraging the growth of the mammary glands, together with a host of other influences.

Birth

Behaviour of the mother after birth has been widely studied (*Rheingold* 1963), but the physiological mechanisms underlying parental behaviour are by no means clear. For example, the behaviour of rodents retrieving young displaced from the nest is easily quantified. Several investigators have shown that the female's retrieving responses is maximal in the latter part of pregnancy and around the time of birth, suggesting a link with the hormones of pregnancy.

Injections of LTH increase retrieving. But hypophysectomized and gonadectomized animals will retrieve young, showing that the physiological basis for this behaviour is more complex than it at first appears. In rats a variety of stimuli, olfactory, auditory, visual, and tactile, are involved in eliciting the retrieving response. *Hamsters* perform a certain amount of retrieving, whatever their physiological condition. On occasion, they may attack or eat the young instead. Careful experimentation with this species shows that the age of young presented in relation to the age of the female's own young is a critical factor.

The effectiveness of the young varies with their age. *Rowell* found that the increasing age of the young is also responsible for the gradual decline in frequency of maternal licking and other changes in maternal behaviour. A mother provides each day with young of a constant age continues to lick them with the same frequency. The same procedure perpetuates milk production; removal of the young curtails it. External stimuli play a particularly large role in maternal behaviour. Nonpregnant rats retrieve freely when they are simply enclosed with young. Very young rodents will retrieve nestmates from an earlier litter, or even from the same one.

Although the predisposition to retrieve is clearly affected by hormones, it may also occur to some extent independently of hormonal control, in the presence of certain external stimuli. Similarly in the production of milk, changes in hormones and other internal factors play a key role. This holds true especially in initiating the general physiological readiness for lactation. But tactile stimulation of the mother's nipples is required for the actual *letdown* of the milk, which is mediated by the release of oxytocin from the neural lobe of the pituitary. Once more the behaviour is a result of interaction between endogenous and exogenous factors.

Hormonal Basis of Breeding Seasons

The onset of the breeding season is heralded by changes in the structure and physiology of the gonads. Either as an endogenous change, or through the influence of external stimuli, the anterior pituitary begins secreting gonadotropic hormones (FSH and LH), which in turn induces growth of the gonads. These hormones also induce the production of the gonadal or sex hormones, mostly androgens in the male and estrogens in the female. The sex hormones elicit changes in structure, modifying the reproductive tract and some of the secondary sex characters, and are the main cause of the onset of reproductive behaviour. The sex hormones are not the only physiological mediators of the onset of the breeding season. The pituitary gonadotropins have direct effects in some cases.

Migratory behaviour, which preludes the breeding season in many species of birds, appears unchanged in castrated finches. Thus, it can occur independently of gonadal hormones. Whether gonadotropins, or some extragonadal sources of androgens and estrogens such as the adrenal cortex, are responsible has yet to be determined. The gonadotropin LH is responsible for the cycling of reproductive and non-reproductive male plumage patterns in weaver finches, and antler growth in deer may be similarly effected. In fishes the thyroid has been implicated in the onset of migration. Spring movement of three-spined sticklebacks, *Gasterosteus aculeatus*, from salt to fresh water occurs normally in castrated animals and can be elicited out of season by thyroxine treatment. In this case the production of the thyrotropic hormone by the anterior pituitary, induced by the lengthening days of spring, is claimed to be responsible for onset of the first phase of the breeding cycle.

Even in these exceptional cases the breeding cycle can hardly proceed far without involvement of the gonads. Arthropods are exceptional in that sexual behaviour and secondary sexual characters seem to be less dependent on direct gonadal control than on activity of the corpora allata. In vertebrates the process of releasing of gonadotropins by the anterior pituitary is initiated in the hypothalamus, which is ideally placed to receive sensory information about external conditions and to respond to endogenous changes within the central nervous system.

Neurosecretory cells become active and their products are transported by the hypothalamo-hypophyseal portal system to the pituitary. These hormones are released into the circulatory system.

Breeding behaviour is then maintained by several complex interactions between gonads, pituitary, and the environment.

Termination of the Breeding Cycle

How the breeding season is brought to an end is not entirely understood. A key role is probably played by the pituitary gonadotropin, LTH. It is implicated in the parental phases toward the end of the breeding cycle, and it is known to cause gonad regression in some male birds. The thyroid gland may also be involved, especially in birds that molt after the breeding season. Whatever, the cause, the pituitary enters a refractory phase at the end of the breeding season. During this phase gonadotropin production cannot be elicited. In time the pituitary again becomes responsive to external stimuli which can again trigger the cycle of secretory activity. The stage is set for the next breeding season.

Sites of Hormone Action

The effectiveness of hormones in eliciting reproductive behaviour is readily illustrated by the standard techniques of removal and replacement therapy. A male rat or guinea pig is castrated. The effect of the operation upon his mating behaviour is determined with standard females in a test situation. After the test he is injected with an androgen preparation which restores the deficit in his mating behaviour. The loss of mating behaviour following castration is even more immediate in females.

In some female mammals estrogen alone suffices to restore estrous behaviour. In other, a combination of estrogen and progesterone is required. Several lines of evidence show that the gonadal hormones are not the only factors involved. Although male hormones injected into females can and normally do induce and sexual behaviour, they may also cause female patterns, especially at high dosages. The same duality emerges when males are injected with ovarian hormones.

Furthermore, behaviour patterns of the opposite sex often appear in the course of natural reproductive behaviour. The mechanisms for male and female behaviour patterns are thus presented in both sexes and are subject to facilitation or inhibition by several factors. Individuality plays a part in the varying effects of hormone administration. Castrated male guinea pigs show individual differences in sexual performance after identical testosterone therapy corresponding to differences they exhibited before castration. *Beach* and *Holz* and *Beach* and *Fowler* also found that the castrate male rats most active sexually before the operation needed the least testosterone to restore their ability to copulate. An element of individuality is also implied by observations that individual female silver foxes and mink are somewhat consistent from year to year in the date of mating.

Mechanisms of Hormone Action

How do gonadal hormones achieve their customary effect? Various lines of evidences show that parts of the central nervous system, particularly in the hypothalamus, respond to the sex hormones by facilitating one pattern of sexual behaviour or the other. Higher centres are also involved. The cerebral cortex is more dispensable for sexual behaviour in female mammals than in males, though it is more important in the parental behaviour of the female. The rhinencephalon and the reticular system are also implicated, and the former in particular seems to have inhibitory effects, so that lesions may cause hypersexuality. And of course the mid-and hindbrain and the spinal cord are intimately concerned with the spatial and temporal organizations of the behaviour patterns used, as they are with all motor activities.

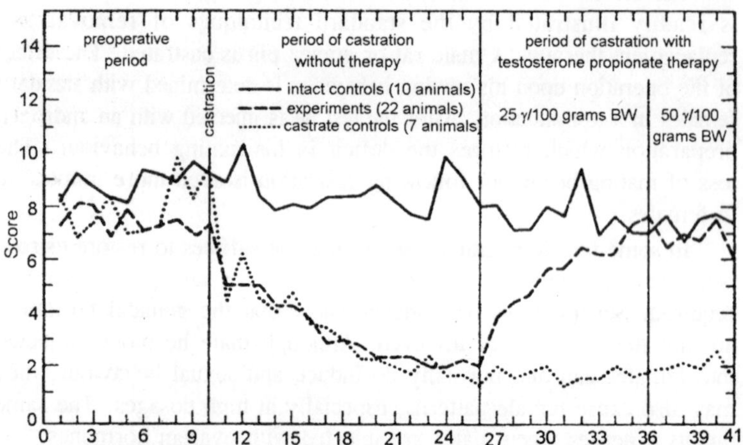

Fig. 7.9. In a test of sexual behaviour, the scores made by male rats declined as a result of castration.

Peripheral Hormonal Effects

Hormones act not only upon the central nervous system to condition or trigger behaviour but also peripherally upon sensory and motor structures. These peripheral effects in turn have direct repercussions on behaviour. Consider, for example, the effect of testosterone on the penis of the male rat. One of the changes that takes place after castration of an adult is a reduction in the development of papillae on the glans penis. Testosterone therapy restores the papillae, and *Beach* and *Levinson* (1950) demonstrated quantitative correspondence between the rate of their reappearance and the onset of renewed mating activity. There are sensory nerve endings immediately below these papillae. It seems reasonable to suppose that their increased sensitivity contributes to the resurgence of sexual activity. The contribution is dispensable, because mounting and other phases of sexual activity can persist after the genitalia are anesthetized, even though intromission and ejaculation are impossible. Peripheral hormonal effects have been invoked to explain other behaviour patterns as well.

Lehrman has shown that in ring doves regurgitation of crop milk to the young, induced by LTH, is in part the result of sensitization of the external wall of the crop to emetic stimuli created by pecking movements of the young. Similarly, the nest-building behaviour of rats is affected by a variety of hormones and other factors, all apparently connected with the threshold of response to temperature change in the environment. Although hypophysectomy increases nest building, it is the relationship with the thyroid rather than the gonads that is significant, because of its role in the control of body temperature. The incubation patch of some birds, induced by the combined effects of gonadal steroids and LTH, may serve as a modified receptor surface with consequent effects on behaviour.

There is greater tactile sensitivity and possible increased sensitivity to the temperature of the eggs and young. Tactile stimulation of the

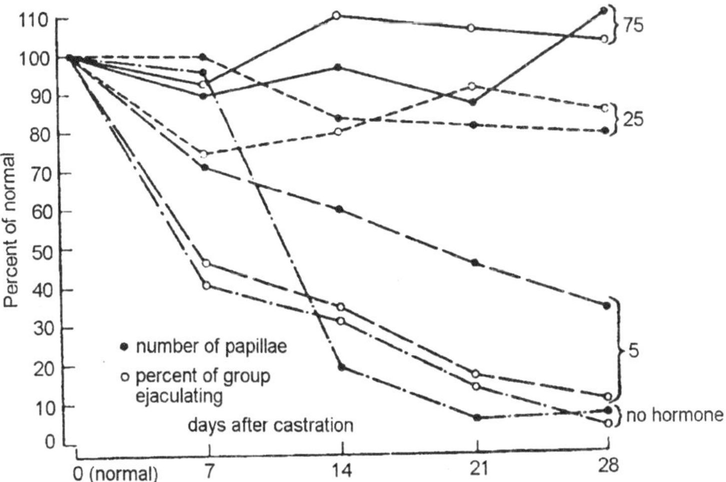

Fig. 7.10. In a test of sexual behaviour, the scores made by male rats declined as a result of castration.

ventral integument may even be needed for complete morphological development of the incubation patch. This sensitivity seems to be related to the timing of changes in the selection of nesting material and mode of construction of the nest. The dichotomy between central and peripheral hormonal effects is not complete. After all, the nervous system in necessarily involved whenever behaviour occurs. Efficient control of the function of many receptors makes the distinction between central and peripheral effects still less absolute. Nevertheless, the theoretical emphasis on peripheral effects by *Lehrman*, *Schnerla*, *Rosenblatt* and *Aronson*, and others has illuminated a previously neglected aspect of the problem. To understand the behaviour completely, the environment in which the animal is placed, as well as its internal physiological state, must be considered.

Variation in the Degree of Hormonal Control

Castration of adult animals generally eliminates sexual behaviour, quickly in females, more gradually in males. The exceptions to this rule have a special theoretical interest. Some men and women continue to engage in sexual behaviour after loss of the gonads. The same is true of various lower primate males, but females generally cease sexual activity after castration. In intact female primates, however, sexual behaviour is not completely restricted to estrus. Male cats and dogs sometimes continue to engage in sexual activity long after castration, but castrated females do not. Moreover, the sexual activity of intact female cats, and indeed of all non-primate female mammals, is restricted to estrus. In rodents castration tends to eliminate the sexual behaviour of both sexes, although there are exceptions.

Beach has suggested a phylogenetic series. Going up the scale there is gradual emancipation of sexual behaviour from complete dependence on gonadal hormones, most evident in the male but perceptible in the female. Thus, series is paralleled by changes in the role of the cerebral Removal of large portions of the cerebral cortex has more effect on the sexual behaviour of male rats that of female rates, but males can still mate after very extensive ablations. In cats there is a similar contrast between males and females. But the sexual behaviour of male cats is more affected by limited cortical ablations than that of male rats. The suggestion is that in some carnivores and primates the cerebral cortex has taken over functions which in lower mammals can be performed only under the influence of gonadal hormones. There is generally a greater involvement of the cerebral cortex in the sexual behaviour of males than of females, irrespective of phylogenetic level. The prominence of cortical involvement, especially in males, suggests that learning may be important in this emancipation from hormonal control. The influence of learning on the sexual behaviour of men has long been recognized. It is also generally agreed that women are less affected in this regard.

Careful experiments by *Rosenblatt* and *Aronson* have established the importance of learning in the sexual behaviour of male cats. Persistence of sexual behaviour after castration is dependent upon prior sexual experience. The possible substitution for gonadal effects of sex hormone secretion from sources other than the gonads, such as the adrenal cortex, has been disposed of effectively. Androgens are necessarily for the initiation of sexual behaviour during ontogeny, however Beach has presented evidence that experience plays a part in shoping the sexual behaviour of other higher mammals as well. The general case is well established for the partial replacement of hormonal control of sexual behaviour by neural control, as a result of experience in certain species of mammals.

Aronson has drawn attention to many variations in sexual behaviour throughout the animal kingdom that remain unexplained. The sexual behaviour of some male fish persists for a time after castration, in others it quickly ceases. Aronson suggests that there may be cases of parallel evolution among teleost fishes, and there is good reason for taking care in deciding exactly what constitutes a phylogenetic series. Certainly experience can play a very crucial role in the development of sexual behaviour of vertebrates at many phylogenetic levels.

8

BIOLOGICAL CLOCK

DEVELOPMENTAL HOMEOSTASIS

Paradoxically, these experiments although they do show the importance of early experience for normal behavioural development, also demonstrate how difficult it is to alter the course of behavioural development through environmental manipulations. In order to produce a rhesus with behavioural disorders, the *Harlows* find it necessary to isolate an infant in a small wire mesh cage at birth or to keep it in a similar environment with just its mother. Neither event has ever occurred under natural conditions for rhesus infants. An abandoned (I.e., isolated) baby would be dead in a short time; and because rhesus monkeys always live in a short time; and because rhesus monkeys always live in groups, a young animal is never entirely alone with his mother for prolonged periods of time, to say nothing of 6 months. Thus no animal in the history of the species has had to cope with the bizarre environmental situations devised by the *Harlows*, and other researchers in this field.

It is not surprising that the animals' developmental systems break down under these conditions. What is surprising and revealing is that otherwise totally isolated infants will develop relatively normal behaviour if they are introduced into a cage with three other infants for just 15 minutes each day. At the outset, the young rhesus monkeys simply cling to one other, but later more time is spent I play. In their natural habitat rhesus babies, beginning at about 1 month, start to play more and more with others about their age. By 6 months they are spending practically every waking moment in the company of their peers. Yet in the laboratory, despite the barren existence of isolated infants, a few moments of playtime a day is sufficient to counteract the effects of extreme depravation.

Rhesus monkeys, therefore, exhibit Developmental Homeostasis, the ability of developmental systems to compensate for environmental (and genetic) deficits and so produce key adaptive traits in animals exposed to less than optimal conditions. Another unexpected demonstration of the resilience of the developmental system of the rhesus is provided by tests of isolated infants reared on a swinging, instead of immobile, surrogate mother. These infants develop more typical behaviour in many regards that those reared on non-moving surrogates. Again, stimulation that only approximates what a body rhesus would receive from its active mother nevertheless provides the basis for the relatively normal development of some behaviour. Exactly, the same points emerge from the Norway rat studies. In order to disrupt normal development, experimenters have had to resort to extremely abnormal environments. No weanling Norway rat has ever been placed in a tiny featureless cage and isolated there fore weeks, prior to the last 100 years.

The rat's genetic program evolved over millions of years in the wild and then modified by selection during domestication, is not able to deal with a situation rats have never confronted. However, even here developmental homeostasis is evident. When the rich-environment rats are given a pale imitation of a typical rat environment (a few other rats to play with and some objects in their cage), they develop more or less normally. Even more remarkable is the quick recovery deprived rats make if given an opportunity to experience anything. When the learning abilities of the two groups were compared on a series of tests, the deprived rats performed more poorly than the others on the first few examinations. But by the end of the battery of problem-solving tests they were doing as well as the enriched-environment rats. The rat's genotype requires a tiny portion of the average experiential input in order to develop basic behavioural abilities. If we were to extrapolate from these studies to human beings, an evolutionary approach would suggest that young human children are probably not high susceptible to various environmental deficits.

The human genetic-developmental system should be as protected against environmental deprivation as a white rat's or a rhesus monkey's developmental program. Yet a large majority of developmental psychologists and the general public, at least in the United States, take it for granted that experience young children have in the first few years of life have an enormous impact on the development of their intelligence, personality, and other aspects of their behavioural repertoire. This view point is reminiscent of the philosophy of the Englishman *John Bocke* as expressed in 1686 in his "*Essay* on Human Understanding." (This paper was written in Amsterdam, *Locke* having been expelled from England 3 years before for unpatriotic behaviour. It was rumored he chose Holland because of a Dutch mistress, but he himself claimed he did so because "there was but little beer in France." Either reason seems sufficient.) *Locke* proposed that the human infant's mind is like a blank sheet of paper on which experience is written and that humans are, in effect, entirely products of what happens to them.

The advice often given to young parents presumes that *Locke's* view is correct and, moreover, that the human infant is highly impressionable and sensitive to its initial experiences in life. There is growing evidence, however, that human genetic programs can compensate for severe environmental deficits even if these occur early in life, as studies of the mental performance of a group of Dutch teenagers have shown. These adolescents were born or conceived at a time when their mothers were being starved during the Nazi transport embargo, which prevented food from reaching the large Dutch cities during the winter of 1944-45. Deaths from starvation were common, and many persons lost one fourth of their body weight.

For most of the famine period the average caloric intake was about 750 calories (roughly one-half the absolute minimum standard) and at one time was as low as 450 calories per day. Those of you who with to visualize the results of 5 months of semi-starvation at a relatively robust 1600 calories per day may examine reference. Pregnant women in large cities were much less than those of the infants born at the same to women living in rural areas and smaller towns in Holland; the towns and countryside were less dependent on food transported to them. One commonly reads that poor nutrition during pregnancy will inevitably damage to the offspring because of the critical nature of this period for the development of the brain.

However, the famine babies did not exhibit a higher incident of mental retardation at age 19 than those born or conceived at the same time in non-famine areas. Nor did these deprived babies score more poorly that relatively well-nourished infants on the Dutch intelligence tests administered to men of draft age. The children were in some way protected from this environmental insult and, despite being born at a lowered weight, suffered no permanent damage. These data indicate that an improper prenatal environment can be compensated for through the resilience of the developmental program. This makes sense. Humans, including pregnant women, have, in all probability, been regularly exposed to food shortages throughout evolution.

Genetic-developmental systems in embryos that were highly sensitive to food deprivation would be likely to produce aberrant individuals with lowered fitness. Those women whose embryonic offspring were relatively insensitive to nutritional deficits would tend to produce normal functional individuals with normal to superior fitness. I am not saying that humans are impervious to all environmental influence or that is a good idea for pregnant women to starve. Still, just because there is an environmental factor that potentially could affect development is no guarantee that it will, in fact, have an irreparable effect. Our genotypes as a general rule can withstand a good deal. *Jerome Kagan*, a child psychologist, recently went to *Gautemala* expecting find that the child-rearing techniques practiced in that country would permanently stunt the intellectual growth of children.

The expectation was based on the knowledge that *Guatemalan* villagers keep young infants from birth to about 1 or 2 years of age in conditions that strike American children treated this *kagan* says that had he seen American children treated this way he would have called the police), the children's developmental program compensates for the first year or so of confinement. By 11 they scores as well on kagan's test as middle-class Americans of the same age. They are socially well-adjusted, capable human beings. If the developing animal really were a passive billiard ball designed to respond to every environmental impact, the end product would often be a disaster. Individuals would run the risk of being permanently warped by a transitory environmental deficit, such as a temporary shortage of food or lack of social stimulation. They might fail to develop the traits that historically have led to survival and reproduction.

The genes of these developmentally "sensitive" individuals would surely be less likely to survive than those that had the capacity to guide development over or around obstacles in a developmental path that had in the past produced animals with high fitness. If behavioural scientists were to look for the adaptive phenomenon of developmental homeostasis in behavioural development, they would probably find many new examples of it. Currently, little is known about how genetic-developmental systems are able to keep on even keel in the face of great genetic and environmental variation. But it seems likely that as more is learned about the regulatory properties of the genotype more will be learned about developmental homeostasis. Perhaps there is considerable redundancy in genetic control mechanisms.

Just as the designers of manned space vehicles provide these machines with a number of backup systems so that one failure is not fatal, so too the genotype may have the capacity to provide developing cells with alternate metabolic pathways for the production or disposal of key substances. Should a mutant gene's product interfere with the manufacture or degradation of a material, other genes may become activated and provide substitute enzymes for these tasks. The maintenance of metabolic homeostasis within cells should promote the regulation of cell function and development. In this way genes can maintain control of the course of development of nervous systems and so steer behavioural development along certain lines. The detailed analysis of the -ole of genetic redundancy and other properties of the genotype that contribute to developmental homeostasis is a formidable but enormously exciting challenge for behavioural scientists.

Circadian Rhythms

Wallace Craig (1918) *opened his classic paper* on appetites and aversions by stating that "the overt behaviour of adult animals occurs largely in rather definite chains and cycles.... " Rhythmicity of behaviour is the dominant theme of his discussion. He culminates with the conclusion that "the active behaviour of the human being is, like that of the bird, a vast system of cycle and epicycles, the longest extending through life, the shortest ones being measured in seconds." The problems of explaining the mechanisms underlying such behavioural cycles are essentially the same whether the period is a year or a fraction of a second.

Annual reproductive cycles are considered by some to be 'exogenous', times by external environmental stimuli. Others feel that are 'endogenous', originating within the animal and thus being to some degree spontaneous (*Marshall* 1960 and *Wolfson* 1960). There are activity cycles associated with the phases of the moon and the tides (*Naylor* 1958, Brown 1959, *Fingerman* 1960, *Hauenschild* 1960, Enright 1963a, b, *Chandrashekaran* 1965. The same questions have been asked about the timing of these rhythms. More progress has been made in the analysis of daily rhythms. More progress has been made in the analysis of daily rhythms, and this chapter we shall concentrate on them.

External Stimuli and Cyclic Behaviour

The role of external stimuli in controlling rhythmical behaviour patterns can be explored in several different ways. Does the cyclic patterning of the behaviour depend on rhythmical external stimuli? Do external stimulus conditions affect the properties of the rhythm? What affects do rhythmical external stimuli have when they are presented, and how does reponsiveness to these external stimuli vary in different phases of the cycle? Do external stimuli partially or completely control the characteristics of an activity sequence--do they time its onset, continuation, and termination and do they underline differences between cycles? Are deviations from the average pattern such about of activity that are longer or shorter than usual, accompanied by changes in respectiveness? We shall try to answer these questions as they relate to activity rhythms associated with the dialy light-dark cycle.

Circadians Rhythms

The behaviour of most animals in some way reflects the cycle of day and night. This influence can be seen in gross locomotor activity, in the patterns of feeding and drinking, in vocalization and in other recurring activities. It may also affect events which occur only once in a lifetime, such as the emergence of an imaginal insect from the pupal case. It is easy to demonstrate these cycles when they recur day after day in the activity of an individual. A mouse cage can be outfitted with a device that recorder all movements. When the records are summarized in the form of a graph with appropriate units for the time axis, the graph reveals a cycle of activity which often reaches a major peak just after sunset and a minor peak around sunrise. There is little activity in the middle of the day.

A diurnal animal such a finch shows a similar pattern, but maximal activity occurs during the day. For the moment let us ignore the detailed internal structure of such a cycle and concentrate on its overall properties. These activity patterns are perfect examples of rhythmical behaviour, recurring regularly with sessions of activity separated by intervals of reduced activity or of no activity at all. What has been revealed by analysis of such daily or circadian rhythms (circa about, *diem*-day, so called because these circles approximate 24 hours under constant conditions)?

Endogenous Pacemaker

One of the critical characteristics of biological rhythms is the existence of an internal self-sustaining pacemaker. What is the evidence

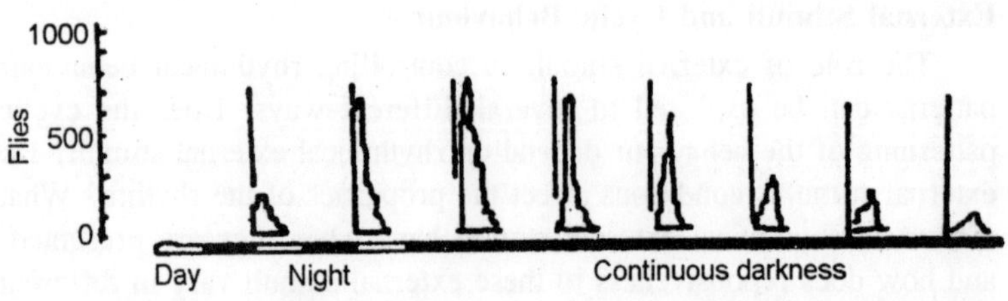

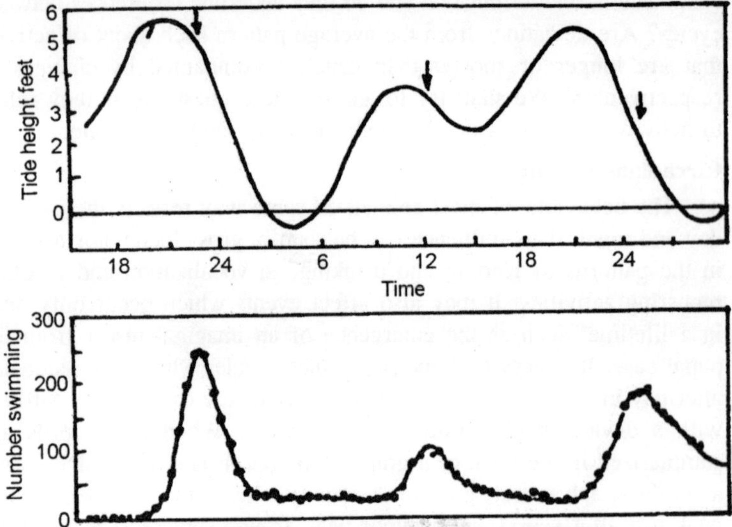

Fig. 8.1. Circadian and tidal rhythms in plants and animals.

for the existence of such endogenous clock mechanisms? As we examine the existence of such internal chronometers bear in mind that what we are measuring are generally the overt, observable manifestations of clock—the periodic changes in physiology and behaviours—not the mechanism itself.

Free-running rhythms

One indirect method of establishing the existence of dome sort of endogenous timer for daily rhythms involves placing an animal in constant environmental conditions, for example, constant darkness. When this is done many organisms exhibit activity rhythms, called free-running rhythms, with a period different from that of any known cyclic environmental variable. The pattern of activity of a flying squirrel (*Glaucomys volans*) housed in constant darkness for several weeks.

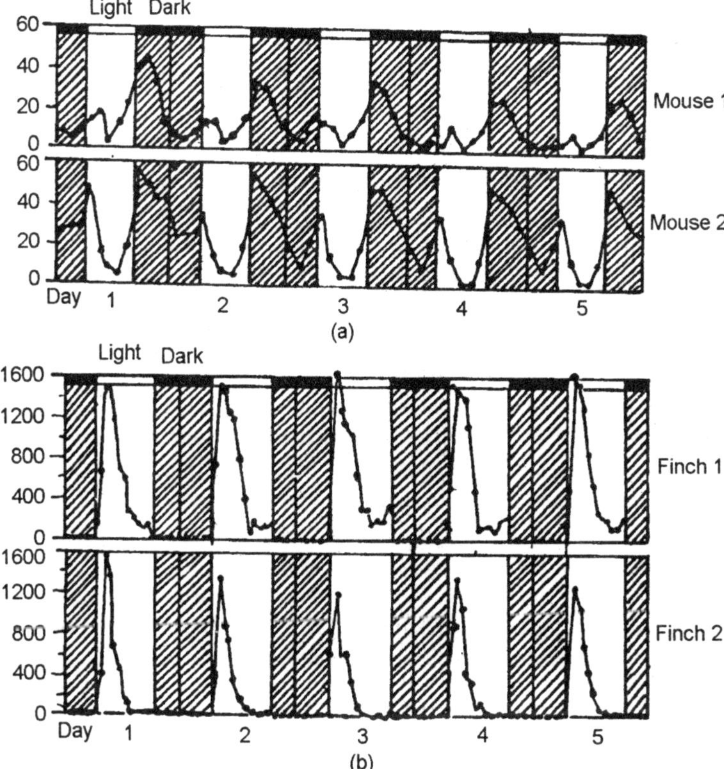

Fig. 8.2. Periodic activity of two mice. Mus musculus (a) and chaffinches, Fringilla coelebs (b) on a 24-hour light-dark schedule in the laboratory.

Since the animal's activity cyclic cue, there is evidence for some endogenously based system of timekeeping. For most species studied, the free-running periods follow what has become known as *Aschoff's Rule* (Aschoff's 1960, 1979). When animals are kept in constant darkness their activity rhythm continues with a period of nearly twenty-four hours, but it *drifts* slightly, becoming somewhat shorter (as in the flying squirrel) or some longer each day.

Aschoff's Rule states that the direction of this drift and the rate of drift away from the twenty-four hour period are a function of the light intensity and of whether the animal is normally diurnal or nocturnal. For nocturnal animals like the flying squirrel, housing under constant dark conditions results in a free-running rhythm with a period of shorter than twenty-four hours: the activity begins slightly earlier each day. Conversely, for a diurnal animal housed in the dark, the free running period is slightly longer each day: the activity begins slightly later each day. There are some exceptions to this rule, but in general the data show a variety of animal species conform to the pattern.

Isolation

Some experiments have clearly ruled out "learning" or other similar influences as a mechanism for biological rhythms. Birds and some reptiles that hatch from eggs can be kept under constant conditions in an incubator from a time prior to hatching until after hatching. If the newly hatched organisms exhibit circadian rhythms, a major component of biological rhythms would appear to be inherited and endogenous. *Haffman's* (1959) studies of lizards have supported the endogenous control hypothesis.

Hoffman maintained lizards eggs under one of three conditions, eighteen-hour days consisting of nine hours of light and twelve hours of dark; twenty-four-hour days, with twelve hours of light and twelve hours of dark; and thirty-six hour days with eighteen hours of light and eighteen hours of dark. Animals from all three groups, hatched and maintained under constant conditions, exhibited free-running activity periods of 23.4 to 23.9 hours. We can therefore conclude that a component of the biological clock mechanism in these lizards appears to be inherited, remains unaffected by various rearing regimes, and is thus endogenous.

Translocation

Additional support for the hypothesis of endogenous control of biological rhythms has come from translocation experiments. Honeybees (*Apis* spp.) are known to visit particular feeding sites at the some time each day. This behaviour is functionally significant because many flowers that provide bees with food are open only at specific times of day. A German scientist, *Renner* (1957, 1959, 1960) working in a specially designed room with constant conditions, trained bees to leave a hive to forage at a specific time each day. He trained the bees in Paris and then flew them, during the night, to New York, where he placed them in a similar enclosed room with constant conditions.

On their first day in New York, the bees began to forage at a time identical to that which would have been expected if they had remained in Paris—that is, twenty-four hours after their last foraging. In related studies, bees that inhabited outdoor hives were translocated from Long Island to California. Investigators found that, although the bees initially foraged at the same time as a the original site, they gradually adjusted to local time by foraging later and later each day. A key aspect of current biological clock models illustrated by these

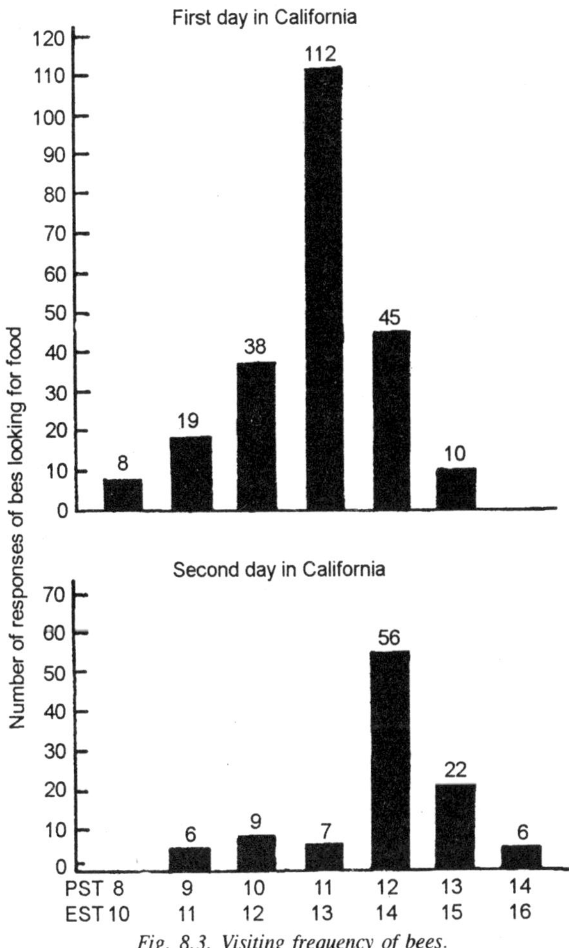

Fig. 8.3. Visiting frequency of bees.

experiments is that while the clocks are endogenous, external conditions can cause the biological rhythms to be reset. Similar relocation studies of circadian skin colour changes in fiddler crabs and shell valve-opening time in oysters revel that these animals, too react initially as if they were still in the original location, but gradually their clocks become reset to local time. The jet lag experienced by humans making long distance airline flights in another example of this phenomenon.

Variation in period

A final source of evidence for the endogenicity of biological rhythms comes from the variation that exists in the natural activity periods of most organisms. For example, measurement of the activity periods of two frogs (*Rana pipiens*) under constant conditions reveals circadian rhythms of twenty-three hours, ten minutes and twenty-four hours, thirty-three minutes. These rhythms do not match the period of any known environmental variable. The animals must have internal chronometers that, due to individual differences, lead to deviations from twenty-four hours rhythms.

Zeitgebers

Many organisms exhibit circadian and circannual periodicities that appear to be closely adjusted to patterns of daylength, temperature fluctuation, or other environmental patterns. The mechanism whereby the period of a rhythm occurs repetitively and coincides approximately with the presence of some external stimulus is called *entrainment*. As we noted in the introduction, cues that provide information to animals about periodicity of environmental variables are called *Zeitgebers* ("time gives"). *Zeitgebers* are the entraining agents defined as those cyclic environmental cues that can enterain free-running, endogenous pacemakers. *Zeitgebers* can influence rhythms by effecting both the phase and the frequency. Returning to the example, when the flying squirrel is provided with a daily cycle of light and dark, its activity rhythm generally conforms to the light cycle—the onset of activity occurs at about the same time each evening.

The squirrel is thus to said to be entrained to the daily light-dark cycle. For most endothermic vertebrates thus far tested, the light-dark cycle is the critical cue for entrainment. Further, in most terrestrial organisms, the daily light-dark cycle is *Zeitgeber*. This is not surprising, since the daily light cycle is usually the most consistent environmental cue in the terrestrial habitat. For animals in the intertidal zone, the ebb and flow of the tides may be a prime *Zeitgeber*.

Ectotherms such as lizards and insects that are unable fully to control their own body temperatures may use temperature or light cues as *Zeitgebers*. When entrained, they exhibit increased activity under warmer conditions and decreased activity during the cooler portions of the daily cycles. Other environmental cues that may serve to entrain some biological rhythms in some organisms are summarized in *Moore-Ede*, *Sulzman,* and *Fuller* (1982). Cycles of food availability, but not water availability, can entrain activity rhythms in several species of small mammals. In other mammals, including humans, social cues may be involved in the entrainment of biological rhythms. Two groups of four volunteer subjects each were isolated in separate, but identical chambers. Each individual maintained an activity rhythms that was synchronous with the others in that room. However, the average free-running periods for the two rooms differed; 2404 hours in one room and 24.1 hours in the other.

To avoid any confounding effect from self-selected light-dark cycles, both groups of subjects were maintained in constant illumination. To further test the possible role of social cues in this synchronization of biological clocks within, but not between, the rooms, one subject was moved from one group to the other. That subject room went rhythm of the new group. The exact nature of the social cues involved in the entrainment of circadian rhythm in humans is not yet known. Social cues may also act in concert with other environmental cues to entrain biological rhythms. When individuals who have just flown across six time zones are required to remain in their hotel rooms, they entrain more slowly to the new local time than do individuals who are permitted to leave their rooms and who thus obtain more social-environmental cues from their new surroundings. If we bring an animal such as a chipmunk (*Tamias striatus*) into the laboratory and with artificial lighting reverse the light and dark portions of the chipmunk's natural cycle, the animal will shift its activity phase to the light portion of the cycle within a few days.

The ability to manipulate the activity phase of animals' daily rhythms has been used by zoological parks to display some nocturnally active animals. By having special quarters that are kept darkened when the zoo is open to visitors and illuminated when the zoo is closed, the public is able to see these animals as active creatures, instead of looking at them only as they sleep. As investigators we can use this type of manipulation to conduct behavioural observations under darkened conditions with nocturnal animals during our own diurnal activity phase. The entrainment of behavioural and physiological rhythms to he cycles of various environmental cues and the phase shifting which takes place when a key environmental parameter is altered are evidence for the *Zeitgeber* component of the system of biological timekeeping.

Model

Using information thus far obtained on biological rhythms we can attempt to depict what is known in the form of a model. The model system accounts for the two major elements known to be important for biological timekeeping, an endogenous self-sustaining pacemaker and a system for entrainment to environmental *Zeitgebers*. The model utilizes light and a circadian rhythm as an example; however, the same model may be used with other types of rhythms, involving entrainment to tidal, annual or other cues.

Cyclic External Stimuli and Circadian Rhythms

The problem of Entrainment

Under constant experimental conditions circadian rhythms do not have periods of exactly 24 hours. Usually the period length is consistently longer or shorter so that an animal originally active at night eventually becomes active during the earth's day. When an animal, which has drifted out of phase with the earth's day, is exposed once more to a rhythm of night and day, it rapidly becomes synchronized again. Its activity cycle becomes locked to the cycle of external conditions and maintains an accurate 24-hour day. There are several questions to be asked about this process of entrainment. What are the effective stimuli?

If entrainment to a 24-hour cycle is possible, what about an 18- or 36-hour cycle' Does entrainment take place in one jump or a series of changes? If the latter, is the degree of change related to the phase relationship between the new stimulus conditions and the original circadian cycle—in other words is there a changing cycle of responsiveness during the existing rhythm? "Entrainment of circadian rhythms is defined as the phenomenon where by a periodic repetition of light and dark (a light cycle) or a periodic temperature cycle, or, more rarely, a periodically repeated stimulus of some other type causes an overt persistent rhythms to become periodic with the same period as the entraining cycle" (*Bruce* 1960). This definition assumes the dominant role of light and temperature as entraining stimuli is experimental situations.

Homeotherms appear to be particularly responsive to 24-hour cycles of light and dark. Poikilotherms such as lizards and insects respond to a cycle of temperature change as well as to light. Activity of the beach isopod, *Excirolana chittoni*, seems to become entrained to the rhythm of the tides as a result of mechanical stimulation by waves on the shore. What happens if we attempt to entrain the rhythm to a period other than 24 hours? It rarely seems possible to entrain animals to day lengths shorter than about 16 hours. The most thorough study of this question comes from the work of *Tribukait* (1956) who subjected domestic mice to light-dark cycles ranging from a 16-hour day to a 29-hour day, with half of each day dark, the other half light. He found that the mice could be entrained to day lengths of from 21 to 27 hours. Activity rhythms failed to synchronize with days longer or shorter

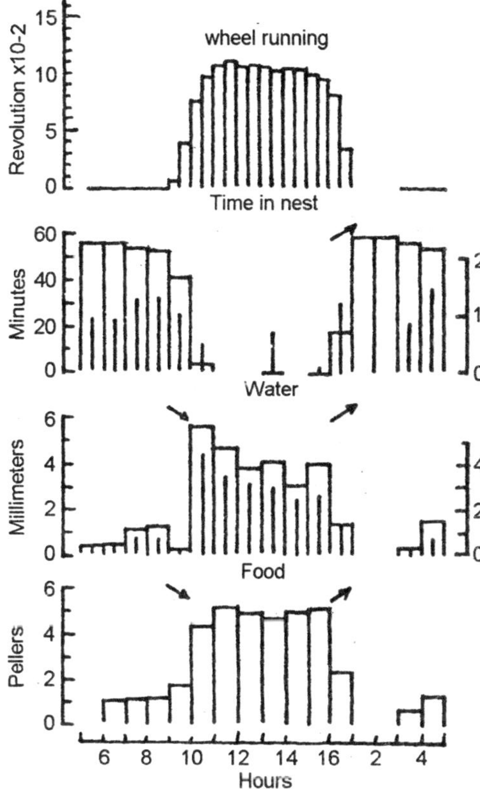

Fig. 8.4. Activity patterns of deer mice, Peromyscum, on a 16-hour artificial day. Artificial twilight transitions are indicated by the slanting arrows.

than this. Instead the mice reverted to an approximately 24-hour cycle. The imposed regime, however, was not totally without effect during these very long and very short days. What was normally one rhythm of activity broke into two. The evening activity peak tended to follow the endogenous 24-hour rhythm, while the morning peak followed the timing imposed by the artificial conditions. Evidently the overall activity rhythm cannot be entrained by a simple on-off lighting schedule if rhythms deviate from 24 hours by more than a certain amount. Similar results have been obtained with hamsters. In contrast, *Kavanau* (1962a, b) has entrained activity, feeding and drinking cycles of deer mice to a 16-hour day (8 light, 8 dark), using artificial twilight transitions rather than simple on-off light control.

More systematic analyses of activity cycles of the same species under different day-night regimes are needed, both with and without stimulated twilight. When an animal is provided with a 24-hour day composed of a series of light-dark cycles, it will often maintain a 24-hour rhythm by responding only to certain cycles. A hamster exposed to a 6-hour rhythm by responding only to every fourth cycle. With a 4-hour day it responds only to every sixth cycle, and so on. Deer mice and cockroaches behave similarly. This phenomenon, called "frequency demultiplication,' demonstrates again the predisposition to respond to stimulus cycles with a period length of about 24 hours. If the problem is approached in a different way by trying to train an animal to perform some activity at a particular time of day, essentially the same result is obtained.

"Even after weeks of training it was not possible to induce bees to look for food, says, every 19 or 48 hours," but they will readily come every 24 hours. These unsuccessful attempts to entrain circadian rhythms suggest a mechanism that restricts entrainment to cycles which approximate 24 hours. Within limits entrainment is clear and accurate Beyond them is difficult or impossible. An apparently endogenous component in circadian rhythms contributes to normal behaviour by restricting responsiveness to rhythmical external stimuli with certain characteristics.

Sensitivity to resetting

Suppose that an animal is exposed to a 24-hour cycle of light and dark which is appreciably out of phase with its activity cycle. Observation of its subsequent behaviour shows that the process of resetting is not sudden. Instead the cycle goes through a series of daily, phase shifts which eventually bring it into line with the pattern of rhythmical stimulation. When synchrony with the external cycle is achieved, phase shifts end. The extent and direction of the phase shifts are found to vary with the phase of stimulation in the exploited the fact that in nocturnal rodents a single brief stimulus of light in otherwise continuous darkness causes a phase shift in the next cycle of activity. The most elegant study of this type is by *De Coursey* (1960, 1961). She interrupted the continuous darkness of flying squirrels with 10-minute periods of light.

The daily cycle of responsiveness was clearly related to the rhythm of locomotor activity. The greatest delay of the next cycle occurred if the light stimulus was presented close to the time of onset of running. The delaying effect declined gradually as the stimulus was presented later in the activity cycle, until several hours after the onset of activity. Beyond this point light stimuli caused an advance of the cycle instead of a delay. Light stimuli had not effect throughout the inactive part of the cycle, until an hour or two before the expected onset of running. A graph reveals the rhythm of changing responsiveness in the course of the activity cycle. Comparable studies of responsiveness to brief light exposure in other species reveal curves with a rather different shape. Maximal phase shifts occur when light stimuli are presented near the time of onset of activity, which in nature is at dusk for a flying squirrel. There is reason to consider sunset the most critical stimulus in setting rhythms of activities for other nocturnal animals.

In diurnal animals dawn is more significant (De Coursey 1961). The effects of a more prolonged light period are likely to be more complicated. The event of resetting may be a compromise between different effects at onset and cessation of the light period. *Brown* (1959, 1960) has pointed the possibility that an animal exposed to constant light may be constantly resetting. If the magnitude of the response to light at the onset of activity differs from that at the end of the cycle, the overall rhythm will show a consistent drift from day to day. In flying squirrels the delaying effect at the start of the cycle is greater than the advancing effect at the end of the cycle. Thus constant light might result in repeated delays in the onset of activity, lengthening the period of the activity cycle.

Moreover, the delaying effect of a light stimulus increases with greater light intensities. This could explain the increase in the period of the cycle with brighter light, which is apparently general in nocturnal animals. It remains to be shown whether the converse effect to constant light in diurnal animals is associated with dominance of the advancing effect over the delay. So the deviation of a circadian rhythm from 24 hours under constant conditions becomes subject to a new interpretation. An endogenous rhythm of motor activity, whatever its period, may be subject to continual resetting by the constant external conditions. Resetting alone cannot explain all deviation from 24 hours in circadian rhythms. Thus *Pittendrigh* (1960) found that hamsters kept on a 23-hour day and then placed under constant conditions had a shorter free-running rhythm than siblings kept on a 25-hour day and then similarly treated. *Harker* (1964) was able to modify the period length of the free-running activity cycle in cockroaches by prior light treatments.

The persistence of such changes under constant conditions cannot readily be related to resetting, unless the prior treatment modifies the resetting mechanism in some durable fashion. Probably there are two kinds of endogenous components, one motor and the other sensory.

Further critical experiments along the lines of DeCoursey's work are needed for an appreciation of the relative contributions of these two types of endogeneity.

Internal Structure of Circadian Activity Rhythms

The manner in which circadian activity rhythms are often presented gives only a gross picture of the actual pattern of activity in time because of the summation of data over long intervals. Presentation of a finer and finer time scale gives a quite different picture. There are actually many short bouts varies somewhat. For example, in mice they may last 106 minutes in the nocturnal phase and 170 minutes in the diurnal phase. The circadian cycle results from variation in the intervals between bouts, which are shortest around dusk and dawn and longest in the middle of the day.

There is in fact another minor cycle much shorter in length imposed on the major one of 24 hours. In rodents, for example, it varies between about 1.5 and 4 hours. There may even be a second rhythm imposed on the first, as indicated in where (a) suggest a 2-hour periodicity, and (c) a periodicity of 4 hours. It is summation of these sub cycles which gives the characteristic bimodal shape to the daily activity curve of mice as it is usually presented. For this reason, the short-term patterning of locomotion, feeding, and other activities is worthy of closer attention from students of circadian rhythms.

Species differences in Circadian activity

The average period of short term cycles of activity varies from species. In shrews *Crowcroft* (1953) has pointed out a correlation with body size. A comparison of daily activity in the large water shrew, the medium-sized common shrew, and the tiny pygmy shrew shows that there are shorter and more frequent bouts of activity in the smaller animals. In these shrews activity seems to be distributed evenly throughout the day. But if the dates are plotted with longer time increments, the same type of bimodal activity curve which occurs in mice emerges. Interestingly enough, the pygmy shrew spends more of its total time resting than the common shrew, in spite of the greater frequency of activity bouts. Variations of this kind lead to species-specific patterns of circadian activity.

Different species of rodents may have strikingly distinct patterns. There are insufficient data available even to guess at the ecological significance of these differences. Although members of the same species generally conform to the same overall pattern of activity, significant individual differences may appear. To explain the survival value of

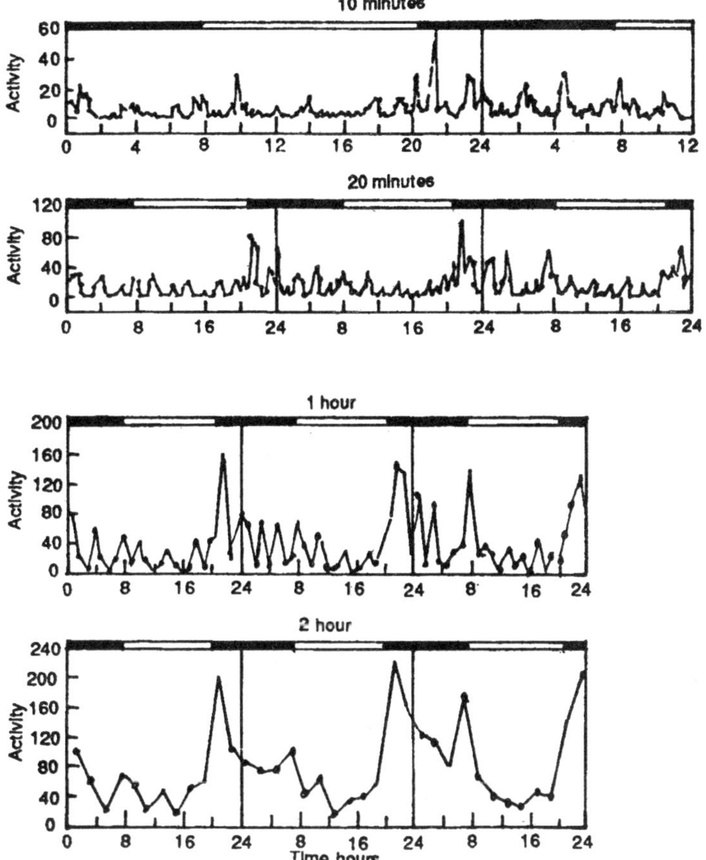

Fig. 8.5. The same activity of a mouse is plotted in increments varying from 10 minutes of 2 hours.

these variations and the mechanisms which underline them, is still another problem.

Seasonal variations

Sometimes there are seasonal changes in the pattern of activity through the day. The winter circadian rhythms of a mouse, *Apodemus sylvaticus*, and the dormouse, *Glis glis*, are lower in amplitude but essentially similar in form to those seen in summer. The mice, *Clethrionomys glareolus* and *Microtus arvalis*, become largely diurnal in winter. The role of internal and External factors in causing these individual, specific and seasonal differences in circadian rhythms is virtually unexplored. The significance of internal factors is revealed by experiments in which the long nights and short days of winter have been simulated to compare their effects with those of shorter nights and short days of winter have been simulated to compare their effects with those of shorter nights and longer days (*Aschoff* 1958, 1960). Mice are active for longer periods of time on winter nights that on summer nights. But in winter, activity starts later in relation to 'sunset' and ceases earlier in relation to 'sunrise' than in summer suggesting that the pattern throughout the year is to some degree independent of the external stimuli of sunrise and sunset.

There is similar seasonal variation of the time of rising of birds. Nevertheless,, effects of changes in the ratio of the periods of light and dark upon activity should not be underestimated. There is also evidence that the intensity of the light under both constant and alternating conditions may have profound effects upon the intensity of activity and upon the ratio of activity to resting. We are still far from any complete explanation of the mechanisms underlying seasonal variations of activity rhythms, however. Nor can we explain the difference between individuals and species and the variations taking place during individuals and species and the variations taking place during individual development.

Mechanism underlying Circadian rhythms

There is little evidence on the physiological identity of mechanisms controlling circadian rhythms. Nevertheless, there is general consensus about the kind of thing to look for. Mary investigators agree that at least two systems which behave differently in some degree must be involved. One is more strictly endogenous, degree must be involve. One is more strictly endogenous, and the other is more responsive to external stimuli. The overall behaviour of the animal is a result of interaction between the two. Brown and his associates were led to this conclusion from their studies of colour change in fiddler crabs. Work on the timing of emergence of the adult fruit flies from the pupae led to a similar hypothesis.

If a culture of *Drosophila* eggs was raised in the dark, the adults finally emerged asynchronously. If a light was presented briefly, subsequent emergence became synchronized. It took place at some multiple of 24 hours after light onset, which would have been dawn, the time of emergence in nature. This periodicity in the behaviour of a population, involving a single event in the life of each individual, has many of the properties of a circadian rhythm and has been explored extensively from this point of view by Pittendrigh and his colleagues.

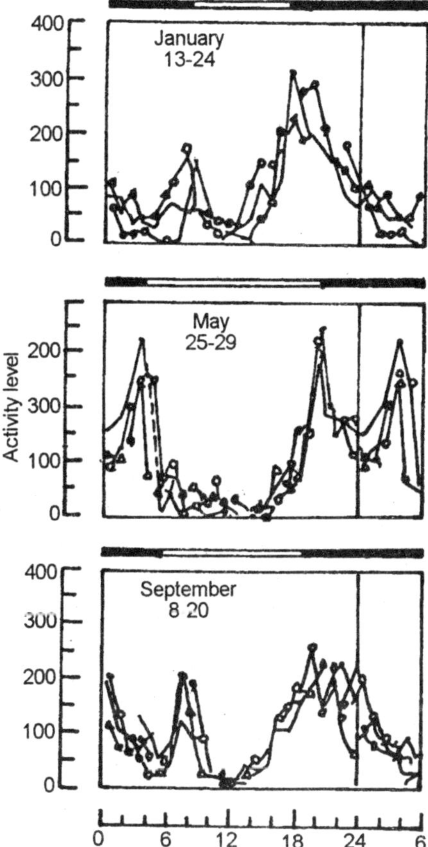

Fig. 8.6. Activity of three mice at different seasons. The black and white bars at the top of each diagram indicate the approximate times of daylight and darkness.

That temperature changes can lead to a temporary resetting of the time of emergence, only to be overridden by the original cycle in the pupae and are still several days from emergency, again suggests that atleast two systems are involved. The experiments of *Tribukait* (1956) on the effects of very long and very short days on the activity peak maintains a cycle of roughly 24 hours which is presumably endogenous, whereas the morning peak keeps time with the artificial cycle.

Here two systems are operating on the behaviour of the same animal at different times. In a general discussion of this problem, *Pittendrigh* (1960) concludes that we must "abandon the common current view tat our problem is to isolate and analyze the endogenous rhythm, or the internal clock," Instead we are faced with the conclusion that "the organism comprises a population of quasi-autonomous oscillatory systems". There are circadian rhythms of human activity under constant conditions (*Kleitman* 1963). Studies of circadian rhythms in human physiology suggest a multiplicity of mechanism. Thus, for a man placed under constant conditions, the periodicities of maximal body temperature and production of water, calcium, and potassium in the urine are not necessarily the same as those of waking and sleeping (*Aschoff* 1965, *Richter* (1965) has received physiological cycles, ranging from minutes to hundreds of days in both normal and diseased human patients. Evidently many cycles can run simultaneously.

Physiological Control of Circadian Rhythms in Cockroaches

Cockroaches show a diurnal cycle of activity which continues with an approximate 24-hour periodicity under constant light and temperature, the exact period length depending on the conditions (Harker 1956. *Roberts* 1960). *Harker's* investigations of the physiological factors underling this behvioural rhythm stemmed from a remarkable discovery. When the suboesophageal ganglion of a cockroach with a certain circadian rhythm ins transplanted into another individual which is arrhythmic, the arrhythmic cockroach takes up the rhythm of the donor (*Harker* 1956, 1960a, b, c). The suboesophageal ganglion thus maintains a circadian rhythm of activity independent of any neural connections for several cycles.

The group of neurosecretory cells in the ganglion responsible for the activity cycles has been located. These cells receive material from neurosecretory bodies, the corpora cardiaca, by way of a small nerve. If this nerve is cut the circadian cycle breaks down, even in a normal light-dark cycle. The phase of cyclic secretory activity in cockroaches is normally set by the alternation of light and dark. According to Harker, this light change is perceived not by the compound eyes but by the simple eyes of ocelli. These ocelli have direct nervous connections to the suboesophageal ganglia, which clearly carry the main responsibility for control of the circadian rhythm. Secretion normally begins at the onset of darkness and activity starts two to four hours later.

However, study of resetting by external stimuli reveals that a second system is also involved. If cockroaches are kept in 12 hours of light and 12 of darkness for several days, resetting can be demonstrated by changing the time of onset of darkness. Immediately after the first sunset at the new time, the suboesophageal ganglion is removed and implanted into an arrhythmic animal which is kept in constant darkness.

In this way, the effects of darkness on the secretory cycle can be compared at different phases of the cycle. There is a definite rhythm of responsiveness. Secretion normally occurs for about 2 hours after sunset. If the new onset of darkness falls in this period there is no resetting. However, if it falls within about 2 hours before or after the secretory period, the periods of 'possible secretion', the cycle is reset.

The next cycle of activity in the implanted animal then begins about 24 hours after the new onset of darkness. At any other time during the period of 'impossible secretion', the onset of darkness has not effect on the secretory rhythm. If the same experiment is done without removal of the ganglia, to observe resetting in the intact animal, there is a striking difference. Onset of darkness is what *Harker* called the period of impossible secretion now causes resetting. This outcome suggests the existence of a second mechanism, also responsive to light-dark changes, that can influence the neurosecretory cycle of the suboesophageal ganglion. *Harker* was able to confirm the existence of this second system by localized chilling of the ganglion *in situ*.

The circadian cycle of cockroaches, as of other insects, can be retarded by chilling at a temperature just above zero. If the whole animal is chilled, all systems should be affected. But by chilling the ganglion alone *Harker* was able to place its cycle of activity out of phase with whatever, other systems the animal might possess. When the chilled ganglion was left in position, the activity cycle was delayed after periods of chilling ranging from 5 to 17 hours. There was no delay after chilling durations of up to 4 to 5 hours. Thus, if the secretory cycle is strongly out of phase with the unchanged second system, it cannot be reset; but if it is only slightly out of phase, with the second system indication onset of secretory activity at a time still within the period of possible secretion, the secretory cycle is reset to its original phase relationship. Complete resetting of the second system by light can be shown by combining chilling and light-dark experiments. The onset of darkness is simulated by covering the ocelli with paint.

In this way the phase relationships can be adjusted so that 4 hours of chilling of the ganglion delay the cycle by 4 hours. This is possible because the second system, modified by the light-dark manipulation, now indicates onset of secretory activity at a time of impossible secretion. *Harker* (1960c) concluded that there are at least two cycles which are light sensitive and which can influence each other. "One of these cycles, refereed to as the secondary cycle, and be immediately reset at any time of day by the onset of darkness, whereas, the other, the neurosecretory cycle, can be reset only when the stimulus occurs at a time fairly close to that when activity would normally occur." The usefulness of these two interacting systems to an animal living in normal field conditions seems to be that they allow for the gradual shift in the timing of the active peak with the change in day length but prevent the resetting of phase by incidental changes of light intensity at abnormal times. A similar explanation applies, but only in the cockroach has such progress been made in locating the structures involved.

9

ORIENTATION AND NAVIGATION

Most animals spend a great deal of time moving around in their environments, but most of them have some sort of nest or "home base" from which they move out regularly to find resources. Such behaviour require that the animal know where it is in relation to its home and how to get back there. Some animals move long distances from their homes and may have different summer and winter homes that are separated from one another by thousands of kilometers. Migration between different ranges requires very sophisticated navigational abilities, but many animals are able to get about with simple mechanisms. The simplest orientation movements are known as taxes (singular : taxis) in which the animal assume a particular spatial relationship to an orienting stimulus.

The nature of the stimulus is denoted by adding the appropriate prefix to the word taxis a *Phototaxis* is movement guided by a light; a geotaxis is movement guided by gravity; a *chemotaxis* is movement guided by detection of some chemical substance. A further distinction can be made between a positive taxis (movement toward the stimulus) and a negative taxis (movement away from the stimulus). The flight of a moth toward a light is a positive phototaxis, whereas the retreat of a cockroach from the same light is a negative phototaxis. Many invertebrates possess a *light-compass reaction*, generally using the sun, in which the angle between the direction of movement and the direction of the stimulus is kept constant. Ants use this type of orientation in moving to and from their nests, but they cannot compensate for movement of the sun. If an ant is held for 2½ hours in a dark box while it is on its way home, the sun will have traveled approximately

37°. When the ant is released, it shifts its direction approximately 37°, maintaining its original angle to the sun.

Orientation to the moon has been demonstrated with intertidal amphipods of the genus *Talitrus* on beaches in Italy. These crustaceans escape seaward when disturbed; populations on costs facing different directions have different but appropriate, escape directions even when tested in the laboratory. On moonless nights or under overcast skies,

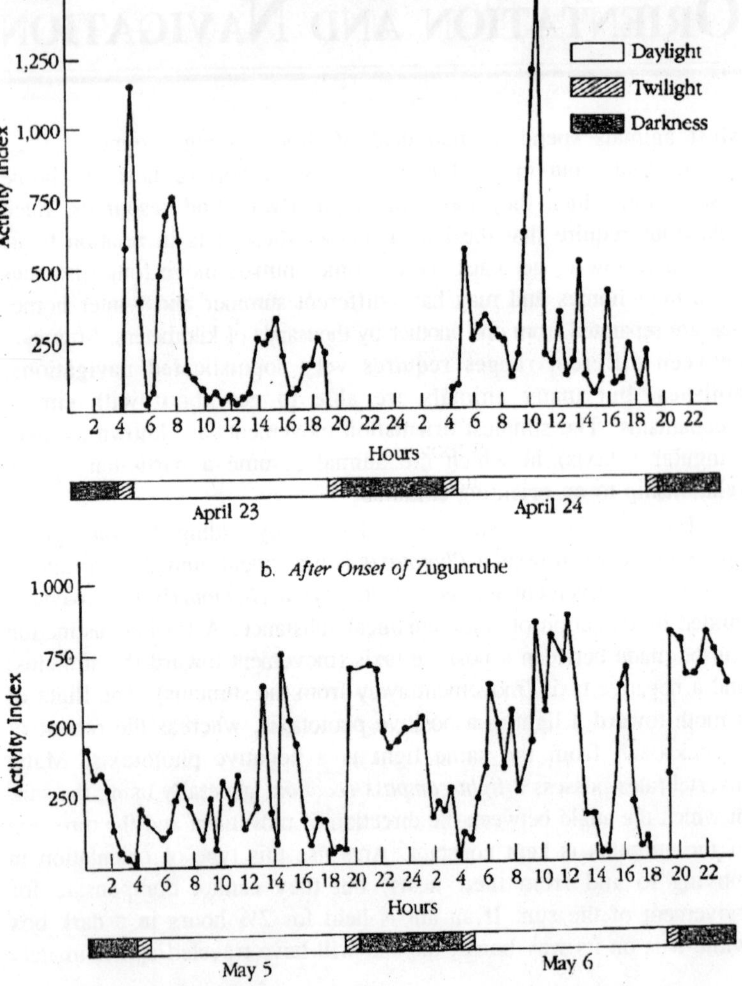

Fig. 9.1. Zugunruhe in male white-crowned sparrow.

this orientation is lost. Unlike ants, however, the amphipods are able to compensate for the movement of the moon and maintain correct compass orientation of their escape. More intensive observations of orientation behaviours have been made with been than with any other invertebrate. Bees can be trained to feed at a particular time at a particular spot in a certain direction from the hive. They have internal clocks that have periodicities of approximately 24 hours (circadian rhythms), which they use to tell time.

Like amphipods, bees compensate for the movement of the sun. If they are held in lightproof boxes at a feeding tray they still fly directly to their hive when released. If trained bees are transplanted to an unfamiliar environment, they still appear at the appropriate direction at the correct time to day to find food. If they are shifted through great longitudinal distances, they first appear at the feeding sites according to the time of day at their training location. Over a period of about after a transoceanic flight.

Navigation

The most spectacular long-distance migrations are undertaken by vertebrates. Pacific salmon hatch from eggs laid in gravel beds of rivers and, after varying amount of time in fresh water—intervals that depend on species and location—they migrate to the ocean, where they feed vigorously until they have reached maturity. At this time they migrate back to the coast, find the mouth of their home river, and swim back upstream, often of the very tributary in which they were born. They then spawn and die. The remarkable migration cycle has two components: navigation at sea to the general vicinity of the home river, ad location of the correct river and tributaries.

Evidence gathered by a number of investigators reveals that salmon and other fishes can orient using a sun compass at sea but that they use a sense of smell to locate their home rivers. Salmon that are transported back downstream successful return to their home tributaries, but they cannot do so if their nostrils are plugged. Even in small lakes, fishes may use a sun compass to orient. White bass (*Roccus chrysops*) in Lake Mendota, Wisconsin, congregate in two small places for spawning. In one study, fish were displaced from the spawning grounds, and floats were attached to them so that they could be followed by observers in a boat. The fish showed good orientation toward the spawning grounds on sunny days, but they swam randomly on cloudy days. The first conclusive evidence of navigational abilities in birds was not obtained until 1951, but for centuries it was known that some birds undertake intercontinental migrations and yet return to specificities to breed. The first studies employed homing pigeons displaced form their lofts.

The pigeons quickly orient homeward under clear skies but depart randomly under cloudy skies. Since the many species, of birds have been shown to use the sun when navigating. Many small birds have migrate at night, however, and they use the stars to orient. Indigo buntings will orient to "stars" in the artificialsky of a planetarium. The physiological condition of buntings may be changed by expositing some birds to increasing day lengths (as in spring and others to decreasing day lengths autumn). In a planetarium, buntings that are in fall condition orient toward the south, showing that their responses to the sky depend upon a physiological star that can be induced simply by manipulating day lengths.

It is also possible to blot out selected constellations in a planetarium to find which ones and necessary for the birds to maintain correct orientation. The relative positions of just a few constellations, particularly those in the vicinity of the Non Star, as used by the to

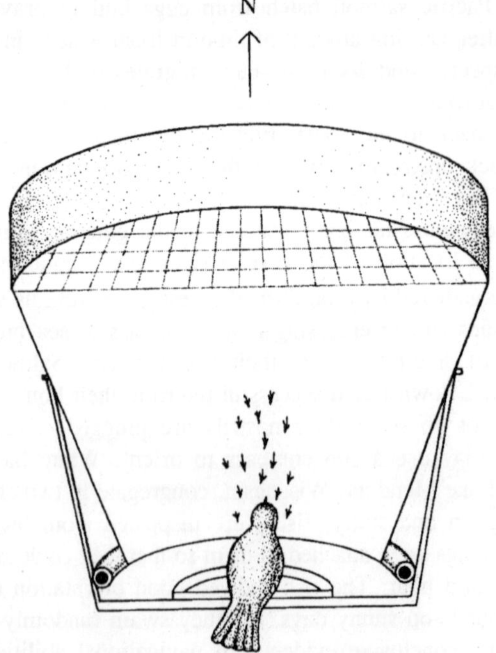

Fig. 9.2. Nightime restleness in a planetarium.

buntings establish their migratory directions; all others can be removed and the birds can still orient correctly. Remarkable though the ability to use the sun and stars to establish directions may be, it still cannot account for true homing. In order to home, bird must know where home is and its own position with respect to home, so that it knows which direction take. It is like a person on a raft lost in the middle of an ocean. A compass is of no use unless the person knows the latitude and longitude of both home and that raft.

Once a bird has determined its position then it can use the sun or stars to take the right direction. Just how birds determine heir positions all unknown, but there is increasing evidence that part of information they use is derived from the earth's magnetic field. Homing pigeons can be disoriented on overcast days by attaching tiny magnets to their necks; the magnets cancel the effects of Earth's magnets field. Also, pigeons are unusually disoriented in the vicinity of sites or *Earth* where there are serious magnetic anomalies due to large non-deposits. There is as yet no magnetic receptor dearly identified, but scientists are convinced that her must be one.

Invertebrates

Migrating locusts, flying in swarms, give the appearance of oriented directional movement. Studies of locust swarms have revealed that wind oxerts a displacing effect on individuals and thus on the swarm over time. Within the swarms there are groups of locusts that are flying in directions different from that of the swarm as a whole. When these subgroups reach the periphery of the swarm, they turn back inwards toward the main flow of the swarm. The means by which honey been (*Apis mellifera*) communicate the location of food sources to one another in the hive and then make foraging flights to locate the food have been studied by a number of investigators.

Bees and some other insects, including ants, can perceive polarized light. The plane of polarization can provide an axis for orientation for these animals. It is also possible that in some animals the perception of polarized light may enable them to locate the position of the sun when it is obscured by partially overcast skies. The use of polarized light for orientation has also been suggested for a variety of other organisms, including fish, octopodes, and salamanders.

Another example is taken from the be killing digger wasp *Philonthus triangulum*. Let us consider first the homing problem of the bee-killing digger wasp *Philanthus triangulum*. Female wasps of this species dig burrows in the sand, to provide a nursery for their offspring. Each female digs several (about 6 or 7) cells off the side of her burrow, and lays an egg in each cell. When the egg hatches into a larva it will need food. That mother wasp goes of her burrow, catches and kills a bee (by stinging), and brings it back to the burrow. She then opens the entrance to the burrow and takes the bee down to one of the cells. She continues to catch and bring back bees, one at a time, until each larva has about two bees to eat. The mother wasp, therefore, does not merely dig a burrow, and later leave it never to return: she departs from and comes back to it many times. On every return she has to find her burrow, distinguishing it from its surrounds.

It is not an easy task to find a particular burrow, because these wasps can nest in quite dense groups; there might be more than twenty burrows within a circle of five yard radius. The problem of how the digger wasp locates her home has a special place in the history of ethology; it was one of the first questions about behaviour mechanism ever to be asked, and experimentally answered. It was studied by *Niko Tinbergen* in 1929, on the heaths and sand dunes of Hulshort in Holland.

Tinbergen first confirmed, by individually marking all the nests and wasps in a particular area, that each wasp does indeed always return to her own burrow. He then performed some simple experiments to test whether the digger wasp recognized her won entrance by the distinctive array of odd objects (haphazardly fallen sticks, pine cones, stones etc.) around it, or by some stimulus emanating from the entrance itself. He placed around the entrance of each of several chosen burrows a neat circle of pine cones. He let them therefore a few days, checking that the wasp still kept returning to their burrows. He Hen waited for the wasps to leave on bee-hunting expeditions, and while they were away he moved the cones a yard or so away from the entrance. When the wasps returned they landed where the entrance 'should' have been, in the centre of the circle of pine cones. Evidently, they recognized the entrance by its surrounding landmarks, not by any stimulus from the entrance itself.

Many of the cues discussed in this chapter with regard to orientation in birds and mammals have also been demonstrated in some invertebrates. Among these are sun-compass orientation in spiders, beetles, and ants; geomagnetic force cues in bees; and landmarks in a variety of species. In addition, data have been reported which suggest beachhoppers (Talitrus) use the moon (lunar orientation) to guide their nocturnal movements.

Fish

We have defined homing behaviour as the use of landmarks and navigation to return to a homesite. An animal may become displaced from its home by weather, or, as in the case of the Pacific salmon, it may return to a home location in the course of the life cycle. Alternatively, researchers may artificially move an organism from its homesite to another location in order to investigate the cues and means of navigation it employs to return home. All species of salmon have similar life cycles. They are laid as eggs in river tributaries all over the northern hemisphere. They live their first few months in the river, then migrate downstream to the sea; they spend two or three years in feeding and growing out at sea, during which they cover thousands of miles; and when mature, they migrate back to exactly the same river, and same tributary, as they were born in. How do they find their way home? The answer is thought to be mainly by smell. The attractive odour may come from either (or both) of two sources: the young salmon which are still in the stream, not yet having migrated to the sea, and any other characteristic odours in the stream.

Salmons have been shown to be capable of the necessary olfactory discrimination, but the most direct evidence that they use their sense of smell comes from experiments, of the kind first performed by *W.J. Wisby* and *A.D. Hasler*, in which the salmon's olfactory sense was impaired. *Wisby* and *Hasler* actually plugged the noses of their salmon, which then homed less accurately than untreated controls recent experiments, the same result has been obtained by cutting the salmon's olfactory nerves. Experiments in which young salmons have been transferred from their stream of birth, to be released in another stream, have been used as evidence that salmon imprint on the smell of their river during a sensitive period just before they migrate downstream.

In one such experiment, *L.R. Donaldson* and *G.E. Allen* took 72,000 young salmon at the 'fingerling' stage when they are about one year old from the Soos Creek Hatchery in Washing and divided them into two groups. In 1952 they released one group (identified by the removal of the right pelvic fin) at Issaquah Hatchery, and the other (which had their left pelvic fin removed) at the University Hatchery. The salmon returned in the winter of 1953/4; they homed on their release sites. None of them returned to Soos Creek, the site of their birth and first year of life, of 71 marked salmon caught at Issaquah Creek, 70 lacked the right pelvic fins; all 124 marked salmon caught at the University Hatchery lacked their left pelvic fins. Results like that suggest that the young salmon learn the odour of their native stream (or of the young salon in it), and later find their way home by seeking that scent. But there is another explanation. Salmon migrate in schools. The results of the transplantation experiment may only be due to the transplanted salmon following the lead of the native salmon.

Recent commentaries, therefore, such as the review by *O.B. Stabell* maintain that the case for imprinting is at best not their sense of smell to guide them home. Whether they learn their home stream's distinctive smell during a discrete sensitive period is undecided.

Table 9.1. Number of Coho Salmon released and later recaptured in two rivers in Washington, with and without their olfactory sense impaired. Salmon with their noses plugged homed less accurately.

Stream of origin		*No. released*	*No. recaptured*	
			Issaquah	*East Fork*
Issaquah	Controls	121	46	0
	Nose plugged	145	39	12
East Fork	Controls	38	8	19
	Nose plugged			

Amphibians

Orientation has been studied experimentally in southern cricket frogs (*Acris gryllus*) and northern cricket frogs (*Acris crepitans*). Experimenters placed captured frogs in a large circular plastic pen in the water, where the frogs could see no landmarks on the horizon. When southern cricket frogs were released in the pen under sunlit skies, they swam in the direction that would have taken them to land had they been released of the shore at their home pond. The results were the same when the experimenters carried the frogs to the pen with or without giving the frogs a view of the sky overhead, when they released the frogs under starlit skies, and when they released the frogs under moonlit skies. When researchers placed the frogs in the pen on a moonless night, with only stars visible in the sky, the frogs oriented themselves in two opposite directions; some headed toward the home shore and others headed directly away from home.

When experimenters released the frogs during the twilight period, after sundown but before the stars or moon were visible, or under cloudy, overcast conditions, the frogs oriented in a random fashion. Cricket frogs held in darkness for periods of thirty hours to seven days exhibited a decline a accuracy of orientation, which culminated in random orientation after seven days, "Dephased" frogs reestablished proper orientation when experimenters exposed them to normal dark/light cycles or to daily fluctuations of temperature and humidity. These frogs appeared to use a learned shore position, celestial cues, and a timing mechanism to orient themselves properly to the home shore. Similar results have been obtained for northern cricket frogs.

Reptiles

Sea turtles perform some of the most remarkable of all migratory feats. They may travel over a thousand miles to a small island to breed. The entire population of the Atlantic ridley turtle, *Lepidochelys*

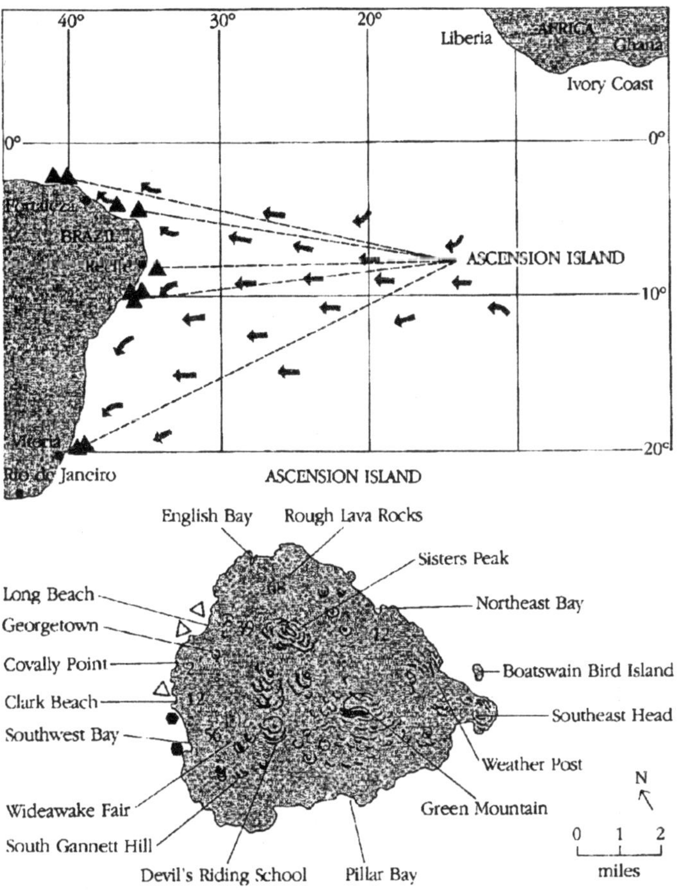

Fig. 9.3. Turtle tagging project on Ascension island.

Kempi apparently breeds on a small stretch of the Gulf Coast of Mexico. So for there is no evidence on the sensory basis of these journeys. *Carr* (1967) has summarized a number of investigations on orientation by green turtles. Female turtles (*Chelonia mydas*) deposit their eggs on sandy beaches on certain islands; they lay about 100 eggs on each trip to the beach, and make from three to seven trips spaced twelve days apart.

In nonreproductive periods the turtles live in areas with abundant turtle grass, generally along the coasts of continental land masses. In a specific study, turtles were banded at Ascension Island. Young turtles that hatches on the island floated on currents that took them to the coast of South America. Adults had to swim against the same current to reach Ascension Island to reproduce. How did they find their way? The current may have certain features that allow the turtles to distinguish it.

Birds

In the last three decades, a variety of studies have been conducted on orientation in and navigation by birds. Many of the hypotheses that have been advanced for birds are also potentially applicable to other animals, such as insects and mammals. Navigation requires bot a compass to provide directional information and a map to provide the animal with information on its position relative to home or some other goal. As the following summaries of evidence for different cues illustrate, considerable evidence exists for the compass, but we still know very little about the "map".

Topographic Features

Use of familiar topographic features is certainly a factor in navigation, but it is probably of secondary importance compared with other cues. Nonetheless, many diurnal and nocturnal migratory birds are influenced by and may utilize topographic features. For example, birds congregate near and fly along coastlines or river valleys, and they may also use landmarks for piloting, particularly in areas at both ends of the migration where they are more familiar with specific topographic features. However, the use of topography as a guidance system has inherent drawbacks. How do first-time migrants who have never learned the landmarks find their way? What if a storm blows birds of the normal route into areas they have never traversed before? Also, visual landmarks alone cannot provide long distances. Both landmarks and certain topographic features the necessary compass direction for proper orientation over could be used in combination with some type of compass mechanism to provide sustained orientation during longer movements. Also, in familiar terrain, landmarks may suffice without any compass.

Sun

The most prominent regular cue for diurnal migratory birds and possibly also for those that migrate at night, is the sun. The latter take a bearing at sunset and use that bearing to fly at night (*Able* 1982). The early studies of bird navigation devoted considerable attention to the sun. *Kramer* (1950, 1952) first showed that European starlings (*Sturnus vulgaris*), placed in an outdoor cage with the sun visible, exhibited migratory restlessness in the appropriate direction in the spring and fall. When Kramer used mirrors to alter the apparent position of the sun, the starlings' migratory restlessness shifted direction in a predict. One way to demonstrate the use of sun compass is to examine whether birds maintain a constant orientation at different times of the day even though the sun's position varies as it traverses the sky. Data from some diurnal migrants provides support for this alternative.

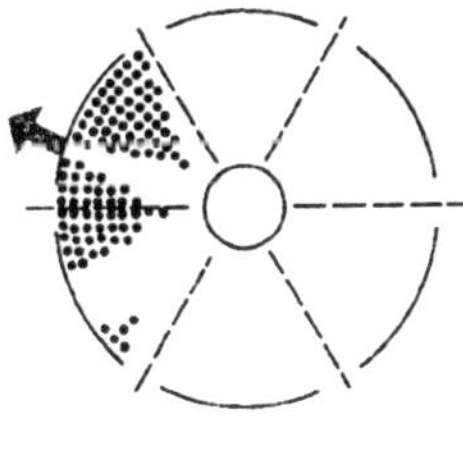

a. Clear skies

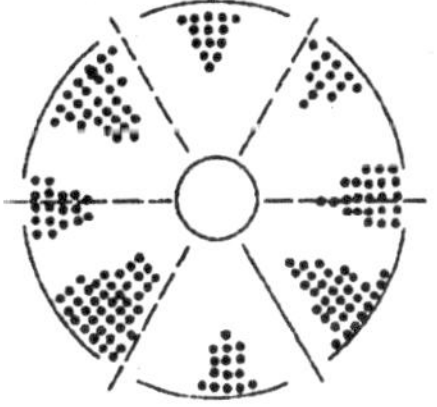

b. Overcast skies

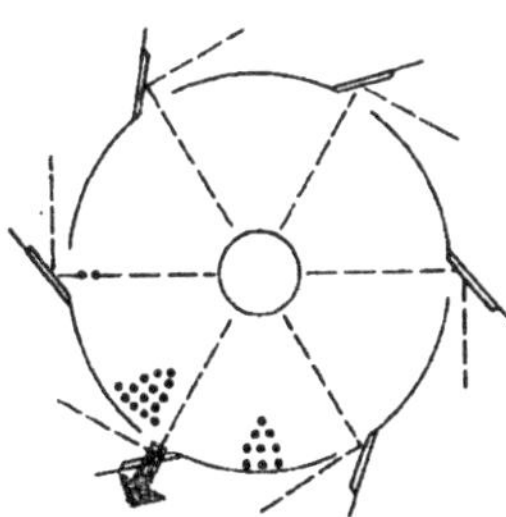

c. Sun's image deflected 90° counterclockwise by mirrors

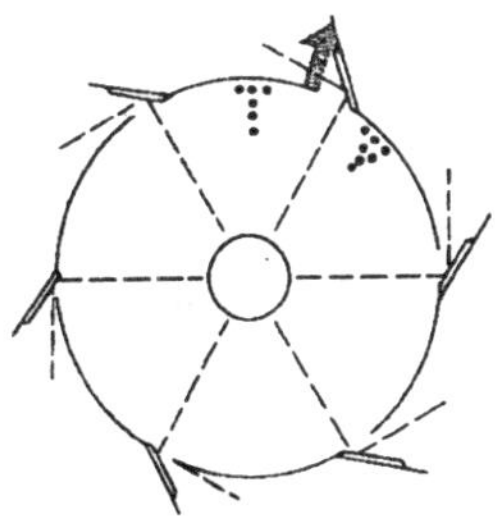

d. Sun's image deflected 90° clockwise by mirrors

Fig. 9.4. Starling orientation experiment.

Clock-shift experiments have provided a second, more rigorous test of sun-compass orientation. First, clock-shifts of six hours, brought about by artificially shifting the birds' internal clock mechanism, resulted in predictable alterations of 90 degrees in the initial bearing of birds heading for the home site. The finding is consistent with the notion that birds use the sun as a simple compass. As a further test birds were clock-shifted six hours birds were using true sun navigation, they should have decided that they were 4,000 miles east of home and headed directly west. However, the birds headed directly east, as would be predicted if the were using the sun *only as a compass* and not as both a map and compass. The sun, then, is an important directional cue for birds that are diurnal migrants and for homing pigeons. Evidence in support of the hypothesis that the sun can be used as both cc¬npass and map is lacking.

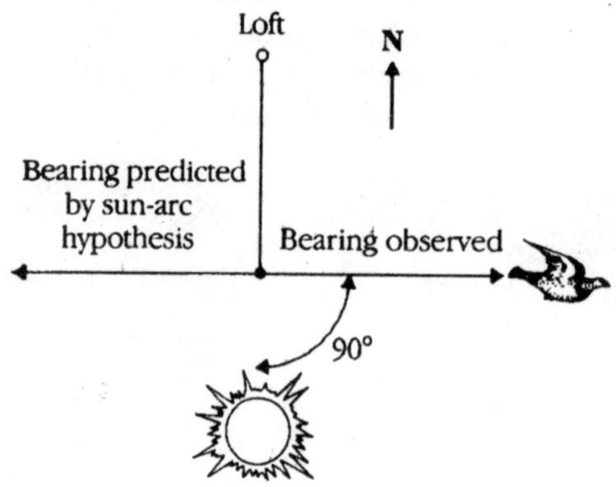

Fig. 9.5. Clock shifted pigeon experiment.

Stellar Cues

When nocturnally migrating birds are placed in outdoor cages with a view of the clear night sky, they exhibit migratory restlessness in a direction appropriate to the seasonal migration. Sauer exposed birds to planetarium skies, which permitted him to manipulate star patterns experimentally. Not only were the results of outdoor studies confirmed, but when he shifted star patterns 180 degrees in the planetarium, the direction of the *Zugunruhe* exhibited by the birds also shifted. Under cloudy skies or with no stars visible and with only diffuse, dim illumination in the planetarium, the birds exhibited random orientation, Sauer's studies on warblers (*Syviidae* have been confirmed in both field and laboratory by *Emlen* (1967a, 1967b) on indigo buntings (*Passerina cyanea*). Emlen's technique is innovative; the bird stands on an ink pad at the bottom of a funnel-shaped cage. The sides of the funnel are covered with paper, and the top is a wire mesh screen that permits the bird to see the overhead sky. When the bird jumps up against the sides of the funnel in its restlessness, it leaves marks on the paper. Investigators can turn the record of these marks into a vector diagram for purposes of analysis.

Meteorological Cues

Migratory activity in a particular region tends to be concentrated on a few days or nights. Birds are capable of responding of favourable weather conditions. Data support the conclusion that birds often fly downward on their migratory flights. Further experiments monitored the natural noctural migratory flights of individuals of several *Passerina* species under various weather and wind conditions. When migrants could observe the sun near the time of sunset or the stars, they flew in the appropriate direction regardless of wind direction.

If observations were made under totally overcast skies with no view of sun or stars possible, birds often flew downwind and thus sometimes in seasonally inappropriate directions. When migrant white crowned sparrows were fitted with forested lenses and released from balloons aloft, they headed downwind, even though that direction was sometimes seasonally inappropriate. These results indicate the primacy of visual cues in birds for determining appropriate migratory direction and that, when deprived of visual cues, some species can determine wind direction.

Olfactory Cues

Several investigators have suggested that olfactory cues, primarily odors carried on the wind, may serve as an aid in orientation, particularly in homing pigeons. The most direct test of such a hypothesis involve the removal of the olfactory capacity and the testing for expected orientation and homing behaviour. When the olfactory sense is blocked or when nerves are transected, behavioural effects that are not related to the homing ability being tested many result; and, therefore, unequivocal evidence of the possible role of olfaction in pigeon homing is difficult to provide. Indeed, studies by *Keeton* and *Brown* (1976), *Keeton*, *Kreithen*, and *Hermayer* (1976), *Schmidt-Koenig* and *Phillips* (1978) and *Hartwick*, *Kiepenheuer*, and *Schmidt-Koenig* (1978) have all supplied date that demonstrate little or no effect on homing behaviour when the nasal passages are blocked or a local anesthetic is used on the nasal epethelium.

Geomagnetic Cues

Evidence from several sources has accumulated that supports a role for geomagnetic cues in orientation and navigation in some birds. After many failures by experimenters in attempting to determine whether birds could even sense the .05 Gauss geomagnetic field of the earth, after a successful laboratory experiment, reported conditioning pigeons to magnetic fields. Measurement of *Zugunruhe* of European robins (*Erithacus rubecuta*) has provided a second line of evidence. Birds were placed in a cage around which a set of Helmholtz coils, which provide an artificial magnetic field, was constructed. *Wiltschko* and *Wiltschko* (1972) have used this apparatus to show that the birds were not responding to the polarity of the horizontal component of the magnetic field; they did not shift the direction of their migratory restlessness when horizontal polarity was shifted. However, they did reverse direction when the vertical component was reversed; and they reversed the direction or orientation in a predictable manner.

In an artificial magnetic field with a zero vertical component and strong horizontal component, the activity was random. Other evidence for the role of magnetic field in orientation come from the work of *Southern* (1969, 1972) on ring-billed gulls (*Larus delawarensis*). Gulls in cages exhibited a strong tendency to walk in a southerly direction except when storms produced temporary aberrations in the earth's magnetic field. Also when gulls with small magnets affixed to their backs were taken some distance form home and then released, they displayed a random pattern of dispersion. Control gulls without the magnets exhibited normal and consistent directional headings to the south.

Additional confirmation has been reported by *Keeton* (1971) ad *Larkin* and *Keeton* (1976), who used bar magnets attached to the wings of homing pigeons. *Walcott*, *Gould* and *Kirschvink* (1979) have reported the existence of an organ, located along the midline of the pigeon's brain, that contains magnetite granules. This organ may prove to be the source of sensitivity to geomagnetism in the pigeon. It is clear from much of the foregoing discussion that many bird species posses multiple systems for orientation; the use of available cues apparently follows some type of hierarchical scheme. Which mechanism the organism employs depends on its preferences for using certain types of cues and the prevailing weather conditions at that location.

Mammals

In their studies on orientation and navigation in mammals, researchers have devoted particular attention to small rodents and bats.

In many of the studies on rodents, experimenters have displace the animal for a distance and observed the rate of return in terms of the number of animals that successfully returned home, the speed with which they returned, and aspects of the rodent's orientation as they left the displacement site. In some studies experimenters have attached reflector tape or radio transmitters to the animals to plot the rodent's course during homing. Rodents have an area, called the home range, in which they carry out their daily functions. When experimenters have determined the home range, they can displace the rodent and monitor its homing behaviour until the rodent's return and recapture within its home range. Data from several species of mice indicate that, when displaced these animals can find their home range again, some from a considerable distance (up to 30 kilometers) and with speeds ranging up to 300 metres/ hour.

In most studies the rodents are generally displaced for distances of only up to 500 metres and may take several days to return to the home range and reenter a trap. The return rate, or the percentage of mice that successfully return home, varies with the species and the nature of the habitat. Overall figures have indicate return rates ranging from 3 to 5 per cent to over 80 per cent. Little work has been done on how rodents orient within their home ranges; however, *Drickamer* and *Stuart* (1984) investigated the movements of deermice (*Peromyscus*). In the winter the patterns of tracks left by the deermice on fresh shown can be mapped. *Drickamer* and *Stuart* used traverses—each segment of the path from tree to tree or from tree to log—to examine

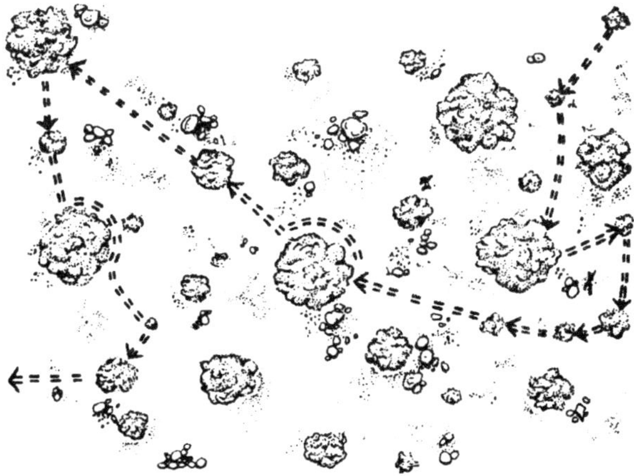

Fig. 9.6. Deermice navigating forest floor.

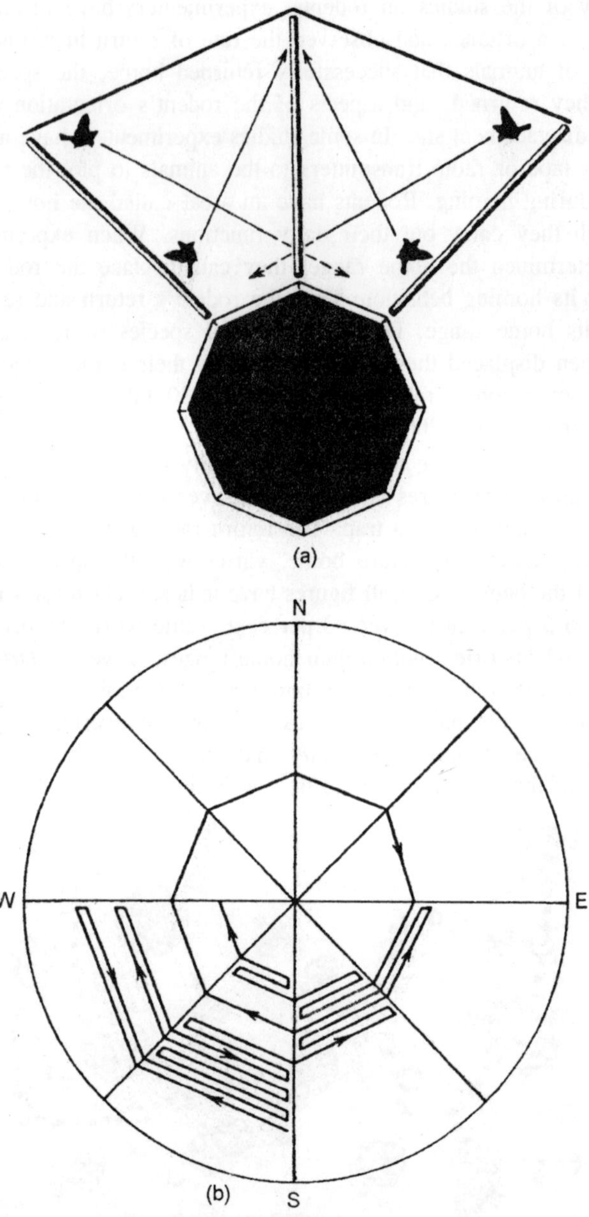

Fig. 9.7. (a) View of the test apparatus used by Merkel and Fromme. The movement from one radial perch to the next is automatically recorded. (b) A sample 30-minute record of the nocturnal movements of robin.

the potential use of cues. They found that 86 per cent of all traverses were oriented from one tree to another tree and covered distances ranging from one to thirty metres.

The concentration of tracks suggests that the vertical tree trunks are used for orienting. A significant positive relationship (r=0.51) between the length of the segment traveled by the mouse and the diameter of the tree to which it was headed further suggests that larger trees serve as more conspicuous objects on the horizon for orientation. Laboratory investigations have supported the hypothesis that deermice may use conspicuous vertical cues on the horizon for orienting to goals. The navigation and orientation systems of bats have received considerable attention. Bats taken from caves, barns, or buildings and displaced some distance often return. Sometimes the distances covered are great. A great deal is known about the homing of bats, but the difficulty of immediately recapturing then when they return and the possibility of missing many that do return make interpretations of the actual homing ability equivocal. Vision and familiar sounds at the home roost may provide terminal information detailing the precise location of the goal. But as in bird homing and migration, mechanisms involved in navigation may be different from those used in piloting. Determination of the initial directional course remains the primary unexplained phenomenon in bat homing.

Preliminary experimental tests of the initial orientation of the fruit bat *Phyllostomus hastatus* seem to demonstrate accurate initial orientation at the release site. The possibility that search patterns provide an explanation of the orientation of migratory flights of bats can be dismissed. Direct migrations covering hundreds of miles have been recorded, and the seasonally appropriate directional flight of red bats, *Lasiurus borealis*, is a commonplace observation in the eastern United States. Yet it is remarkable that no author has suggested any explanation for how these travelers find their way! Bats often collide with towers in the same circumstances and often on the same night as migrating birds do. Perhaps celestial orienting cues are a source of directional information for them.

Other Navigation Mechanisms

Orientation without Optical Cues: an Artifact? Several authors have reported oriented migratory fluttering in birds in the apparent absence of directional optical cues. *Merkel* and *Fromme* (1958) and *Fromme* (1961) worked with the European robin, *Erithacus rubecula*, and the whitethroat *Sylvia communis*. Birds were evenly and faintly

Fig. 9.8. Orientation of the robin, Tilthacus rubecala, in experimental Chamber lacking directional visual cues.

illuminated, but they had no view of the sky. Various experimental locations were used, the most restricted of these being the placement of the test apparatus in a cellar without windows. Here directional orientation persisted, even in a steel cylinder with airtight doors, as long as one of the doors remained open. The magnetic field of the earth at the cage was undisturbed. This finding contrasts with Sauer's demonstration that garden warblers became disoriented when placed in a diffusely lit planetarium or under natural overcast.

The conditions and recording methods used by *Merkel* and *Fromme* and different from those reported by *Sauer,* so that no direct comparison of results is possible. *Precht* (1961) and *Gerdes* (1961, 1962) recorded the escape behaviour of black-headed gulls, *Lerus ridibundus*, displaced from their nests. These birds also maintained a significant orientation, heading toward home not only under overcase conditions but also indoor rooms. In all these experiments in which orientation has been demonstrated under apparently cueless conditions, the birds have not been as highly oriented as birds able to view the everhead sky. The problem was reinvestigated by *Perdeck* (1963), again with the European robin. *Perdeck* concentrated on methodological features. He showed that orientation was random under indoor conditions as long as the orientation device was rotated. When the device remained fixed during experiments, directional preferences not related to the migratory orientation of the species were expressed. In addition, *Perdeck* discovered that the presence of other robins in the same building provided directional auditory cues which influenced the test chamber responses.

10

SEXUAL BEHAVIOUR

Biological fitness is a function not only to an animal's ability to survive but also of its ability to contribute to the gene pool of the next generation and subsequent generations–to reproduce. An organism that copes successfully with its problems of obtaining food and water, finding shelter, and avoiding predators, yet fails to reproduce, still may have a fitness of zero. Reproductive behaviour is obvious importance to fitness and lies very close to the very definition of species. Because of its obvious importance, and because animal behaviourists generally treat reproductive behaviour as neither individual nor social behaviour reproductive behaviour is being taken under a separate head. Reproductive behaviour posses a set of related questions about mating and the rearing of offspring. Before conceptions, there maybe a struggle among males, and a characteristic (often strange) sequence of behaviour patterns in which the sexes court each other. The term courtship, as used by the ethologists, refers to all the behavioural interactions of the male and female which come before, and lead up to the fertilization of eggs by sperm. Its form and ostentation vary among species. In some species it does not even exist. I others, such as the stickleback (which we shall consider below) it lasts a few minute.

In others it may last for months. Male and female waved albatrosses (*Diomede irrorata*), which live on Isla Espanola of the Galapagos Islands, may court each other, with an extensive repertory of stereotyped movements of the neck and bill, for several hours a day, day in day out for much of the year. The question courtship poses is why it exists, and why, in some species, it has taken on so extravagant a form. One answer is that courtship ensures that animals mate with other individual, of the correct species, sex and condition. Its extreme development, however, of numerous energetic and colourful displays, and extreme structural modification in males, cannot be so easily explained.

It is difficult to believe that all the paraphernalia of sex is needed merely to ensure that males are not confused with females or males of another species: that could be achieved much more quickly and less colourfully. Moreover, the most bizarre sexual characteristics, such as the amazing plumage of male birds of paradise, probably decrease the chances of survival of their bearers. They use up energy, hamper flight, attract predators. They are therefore something of a puzzle. They must possess some hidden function to compensate their obvious disadvantages, because if they did not they would have been eliminated by negative selection. Darwin invented a special theory to solve the problem, his theory of sexual selection.

Choosing a Member of the Right Species

Matings between members of different species are very rare in nature. For example, in a sample of 725 copulating pairs of two species of cicada, which are distinguishable but similar in appearance to the human eye, only seven contained members of two species; the other 718 were matings of the same species. Different species often do not get a chance to interbreed because they live in different geographic areas, or in difference localities within an area (one species at the tops of tress, the other on the ground for example), or they might be active at different times of day. Of the few species that do have the opportunity to interbreed, none do so frequently. They usually do not because the one species does not respond to the sexual lures of the other. It is adaptive for animals not to mate with members of other species.

Hybrid offspring are usually inferior to those produces by matings of members of the same species, often because they are sterile. The best known example of a sterile hybrid is the mule. The mule is the offspring of a he-ass and a mare; if a she-ass mates with a stallion the result is a 'hinney', which is also sterile. Natural selection will act to prevent animals from mating with members of other species if the offspring so produced would be sterile. Natural selection will favour animals that produce normal healthy offspring by choosing to mate with members of their own species, rather than producing sterile hybrid offspring. There have been many investigations of the factors animals use to ensure that they mate with members of the same species. Let us take crickets as an example. There are approximately 3000 species of cricket in the world. They do dot all live in the same place, but in any one place there may be more than one, and perhaps half a dozen, species. They can tell each other a part by their distinctive songs. The songs are produced by males when they rub their wings together, and are heard by the female through the ears on her front legs; they are familiar to us from night-time walks, or the vicarious experience of the cinema.

The chirrups of the crickets are calling male. Females are attracted only by the songs, as can be demonstrated by putting out a loudspeaker broadcasting the tape-recording song of a male cricket: females of that species will approach it. Females, moreover, only approach loudspeakers playing songs of their own species. They are therefore using the song to choose a mate of the correct species. (Parasitic flies are also attracted to the loudspeaker. Parasitic flies are one of the hazard of the male cricket's sex life; they are attracted to calling males, on who they lay their eggs. The growing parasitic larvae soon inactivate the male. It is a common risk in many species that by broadcasting to females, a male attracts predators and other enemies). The all-importance of song in the species recognition of crickets can be shown by another experiment. Female crickets can be trained to walk along a 'Y-maze' on which they have to turn to the left or right. If, for instance, the loudspeaker on the left were playing the song of the female's own species, while the one on the right played the song of some other species, the female turns to the left. The side of the songs can be reversed to control for any bias the female may have for turning in one direction or the other. The Y-maze has also been used to see how hybrid female crickets behave. If a male and female cricket of two closely related species are forced to mate, some hybrid offspring will be produced When *R. Hov* put these hybrid females on a Y-maze he found they preferred the songs of hybrid males to those of either parental species.

The result is interesting because it bears on the following problem. When a new cricket species evolves, the male song and the female receptor must change in a co-ordinated fashion: a change in one without a change in the other would be disastrous. The fact that the song and the receptor change together in the hybrids suggests that there is some control mechanism which prevents the two from becoming uncoupled from each other; such a mechanism would have the effect of making it more likely that, whenever a male or female cricket changed its song or song-preference during evolution, some other individual of the other sex will have made the complementary change.

Sexual Selection

The success of an individual is measured not only by the number of offspring it leaves but also by the quality of probable reproductive success of those offspring. Thus, it becomes important who its mate will be. *Darwin* (1871) introduced the concept of *sexual selection*, a special process that produces anatomical and behavioural traits that affect an individual's ability to acquire mates. Sexual selection can be divided into two types: one in which members of one sex choose certain mates of the other sex (*intersexual selection*), and a second in which individuals of one sex compete among themselves for access to the other sex (*intrasexual selection*). One result of either type is that the sexes come to be different–that is, *dimorphic*.

Intersexual Selection

Intersexual selection individuals of one sex (usually the males) "advertise" that they are worthy of an investment; then members of the other sex (usually the females) choose among them. Most naturalist after *Darwin* discounted the importance of mate choice in evolution, but it has recently become a popular topic of study. *Fisher* (1958) used birds as an example. Suppose a plumage characteristic in males is linked to some survival or reproductive advantage for the bearer. If females prefer those males as mates, they will have higher reproductive success and will likely produce sons with the trait and daughters that prefer the trait when choosing a mate of their own. Further development of the trait will proceed in males, as will the preference for that trait will proceed in males, as will the preference for that trait in the females, resulting in a runaway process.

A possible result of such a process is seen in the peacock. Sexually selected traits may become so exaggerated that survival of the males is reduced. Counterselection in favour of less ornamented males will occur—due, for example, to greater predation on the more ornamented males–and sexual selection is brought to a steady state. As an example of intrasexual selection in action, consider the bowerbirds of southeast Asia. Bowerbirds are unique in that they build a structure whose only function seems to be to attract mates. The male satin bowerbird (*Ptilonorhynchus violaceus*) of Australia ad New Guinea Stations itself alone in the forest, clears a space, weaves the bower from twigs, and decorates it will brightly coloured objects *Gilliard* (1963, 1969) reasoned that bowers function in place of showy plumage, and noted that those species with the fanciest bowers had the drabest plumage. He argues that such behaviour arose to synchronize mating between males and females.

Intrasexual Selection

Intrasexual selection involves competition with one sex (usually males), with the winner gaining access to the opposite sex. Competition may take place prior to mating, as with ungulates such as deer (family Cervidae) and antelope (family Bovidae). Typically, males live most of the year in all-male herds; as the breeding season approaches, males engage in highly ritualized battles, using their antlers or horns. The winners of these battles gain dominance and do most of the mating. Antlers are better developed in those cervid species where males compete strongly for large groups of females (*Clutton Brock, Guiness, and Albon* 1982). It is often difficult to determine which type of sexual selection is operating to produce an observed effect, since members of both sexes may be present during courtship.

The "roughout display" of the male brown-headed cowbird (*Molothrus ater*) consists of a "glunk-chee" sound, made as the bird spreads its tail, lowers its wing, topples forward, and nearly falls over. This display seems to intimidate other males and to attract females. Similarly, the chirp of male field crickets (family Gryllidae) is used to exclude other males from an area and to attract females. Females may incite competition among males and thus may gain some control over the choice of mate. For example, female elephant seals (*Mirounga angustirostris*) vocalize loudly whenever, a male attempts to copulate. This behaviour attracts other males and tests the dominance of the male attempting to mate. In response to the female's sounds, the dominant harem master will drive off low-ranking, potentially inferior mating parents. Males may fight directly at the time of copulation, as exemplified by the male of the yellow dung fly (*Scatophaga stercoraria*).

A male will sometimes attack a mated pair, displacing the male and mating with the female. Nor does competition among male to sire offspring cease with the act of copulation. Females of many species store sperms and may remate before sperms from the previous mating are used up, creating the possibility of sperm competition (*Parker* 1970b). Sperm competition can be thought of as a selection pressure leading to two opposing types of adaptation in males: those that reduce the chances that a second male's sperm will be used (first male advantage), versus those that reduce the chances that the previous male's sperm will be used (remator male advantage). First male adaptations include mate-guarding behaviour and deposition of copulatory plugs, both of which reduce the chance of sperm displacement by a second male.

In fruit flies (*Drosophila melanogaster*) first males may also transfer to females "anti-aphrodisiac" substances that inhibit courtship by other males (*Jallon, Antony and Benamar* 1981). Female spiders store sperm for long periods and often mate with several males. In laboratory studies of the bowl and doily spider (*Frontinella pyramitela*), *Austad* (1982) found that the first male's sperm had priority in fertilizing eggs over sperm of subsequent males, as is the usual case in spiders. Among insects, however, the advantage usually seems to accrue to the sperm of the remator male. In the yellow dung flies he studied, the last male to mate sired 80 percent of the offspring subsequently produced.

The ultimate remator made adaptation may be the penis of the demselfly (*Calopteryx maculata*) which has a dual function: it removes sperm deposited in the female by a previous male via a special "sperm scoop" and then replaces it with its own. Moths and butterflies (Lepidoptera) produce two kinds of sperms. One kid is the usual type (*eupyrene*) that fertilizes the eggs. The second type (*apyrene*) contains no nuclear material but may comprise more than half the sperm complement. Why should males waste energy on these "dud" sperm? *Silberglied*, *Shepherd*, and *Dickinson* (1984) suggest that they are the result of sperm competition, possibly displacing eupyrene sperm from first males or delaying remating by the female. Following conception, male-male competition may take a different form.

In mice the *Bruce effect* operates early in pregnancy: a strange male (or his odor) causes the female to abort and become receptive (*Bruce* 1966). Among langur monkeys (*Presbytis entellus*) strange males may take over a group, driving out the resident male. The new male may then kill the young sired by the previous male. Females who have lost their young soon become sexually receptive, and the new resident male can inseminate them. Infanticide by adult males thus may be viewed as a "remator male" adaptation.

Timing or Seasonality

Most of animals breed in a particular time of their life or season, that is why they are usually known as seasonal breeders. In lower animals the timing or seasonality is less understood. Timing in the higher animals is a more complicated affair. Something is known of fish, birds, and mammals of the Northern temperate zone. Reproduction of most of these begins in spring. The first phase is migration towards the breeding grounds. This is done by all individuals at approximately the same time, though there may be weeks between the arrival of the first and the last comer. This rough timing is again due not to social behaviour but to reactions to an outside factors. The main factor here is the gradual lengthening of the day in late winter. Various mammals, birds, and fish have been subjected to artificial day-lengthening.

The result was that the pituitary glad in the brain began to secret a hormone which is its turn affected the growth of the sex glands. These then began to secret sex hormones, and the action of these sex hormones on the central nervous system brought about the first reproductive behaviour pattern, migration. Often a rise in the temperature of the environment as an additional effect. In some cases, the timing process is not very accurate. The different individuals do not all react to the lengthening of the day with the same promptness. There may be a considerable difference between the male and the female of a pair. It has been found, in pigeons and in other animals, that if the male is further advanced than the female, his persistent courtship may speed up the female's development. This has been found in the following way. When a male and a female are kept separately in adjoining cages so that they can see and even tough each other, but

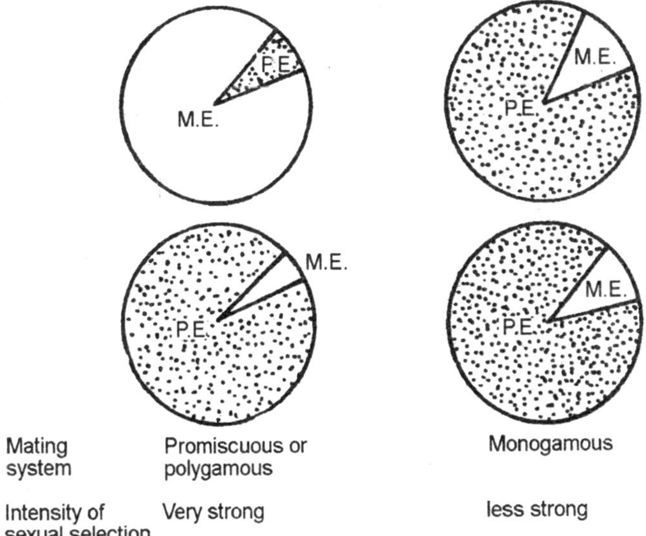

Fig. 10.1. The total resources of time and energy used by an animal in reproduction is referred to as reproductive effort.

are prevented from copulating, the persistent courtship of the male will finally induce the female to lay eggs. These of course are infertile.

It may occur in captivity, when no males are available, that two female pigeons form a pair. Of these two, one then shows all the behaviour normally shown by the male. And although their reproductive rhythms may have been out of step at the beginning, the final result is that they both lays eggs at the same time. Somehow, their mutual behaviour must have produced synchronization, not merely of behaviour but also of the development of eggs in the ovary.

The Sex Ratio

If one male can fertilize the eggs of dozens of females why not produce a sex ratio of, say, one male for every 20 females? With this ratio the reproductive success of the population would be higher than with a 1:1 ratio since there would be more eggs around to fertilize. Yet in nature the ratio is usually very close to 1:1 even when males do nothing but fertilize the female. The adaptive value of traits should not be viewed as being 'for the good of the population', but 'for the good of the individual', or more precisely 'for the good of the gene;. As *R.A. Fisher* first realized the 1:1 sex ratio can readily be explained in terms of selection acting on the individual; his argument is simple but subtle. Suppose a population contained 20 females for every male.

Every male has 20 times the expected reproductive success of a female (because there are on average 20 males per male) and therefore a parent whose children are exclusively sons can expect to have almost 20 times the number of grandchildren produced by a parent with mainly female offspring. A female biassed sex ratio is therefore not evolutionarily stable because a gene which causes parents to bias the sex ratio of their offspring towards males would rapidly spread, and the sex ratio will gradually shift towards a greater proportion of males than the initial 1 in 20. But now imagine the converse. If males are 20 times as common as females a parent producing only daughters will be at an advantage. Since one sperm fertilizes each egg. Only one in every 20 males can contribute genes to any individual offspring and females therefore have twenty times the average reproductive success of a male. So a male biassed sex ratio is not stable either. The conclusion is that the rarer sex always has an advantage, and parents which concentrate on producing offspring of the rare sex will therefore be favoured by selection.

Only when the sex ratio is exactly 1:1 will the expected success of a male and a female be equal and the population stable. Even a tiny bias favours the rarer sex: in a population of 51 females and 49 males where each female has one child, an average male was 51/49 children. This average value is the same whether one male does most of the fathering or whether fatherhood is spread equally among the males.

Table 10.1. In polygamous or promiscuous species males have a much higher potential reproductive rate than females.

Species	*Maximum number of offspring Produced during lifetime*	
	Male	*Female*
Elephant seal	100	8
Red deer	24	14
Man	888	69
Kittiwake gull	26	28

The argument that the sex ratio should be 1:1 can be refined by re-phrasing it in terms of investment. Suppose sons are twice as costly as daughters to produce because for example, they are twice as big and need twice as much food during development. When the sex ratio is 1:1 as son has the same average number children a daughter. But since sons are twice as costly to make they are a bad investment for a parent: each of its grandchildren produced by a son is twice as costly as one produced by a daughter. It would therefore pay the parents to concentrate on making daughters.

As the sex ratio swings towards a female bias, the expected reproductive success of a son goes up until at a ratio of two females to every male an average son produces twice the number of children produced by an average daughter. At this point sons and daughters give exactly the same return per unit investment: a son costs twice as much to make but yields twice the return. This means that when sons and daughters cost different amounts to make, the stable strategy in evolution is for the parent to *invest* equally in the two *sexes* and not to produce equal numbers. An example to illustrate this point is *Bob Metcalf's* (1980) study of the sex ratio of two species of wasp; *Polistes meticus* and *P. Variatus*. In the former females are smaller than males, while in the latter they are similar in size. As predicated, the *population* sex ratio is biassed in *P. metricus* and not in *P. variatus*. In both species the investment ratio is 1:1. The prediction that parents should invest equally in sons and daughters does not always hold, and demonstrations of these deviations from 1:1 investment are among the most convincing pieces of evidence that the sex ratio has evolved in the way suggested by Fisher. We will pick out some example in the following paragraphs.

Local Mate Competition

Fisher's theory predicts a different outcome when brothers complete with each other for mates (so-called '*local mate competition*'). Suppose, for example, that two sons have only one chance to mate and that they compete for the same female. Only one of them can be successful in mating, so from their mother's point of view one of them is *wasted*. This is an extreme example, but it illustrates the general point that when sons compete for mates their value to their mother is reduced. The mother should therefore bias her ratio of investment towards daughters. The exact degree of bias predicted by Fisher's theory depends on the degree of local mate competition. Extreme competition is to be expected in species with limited power of dispersal (because brothers will stay together in the same place) and therefore, it tends to be associated with inbreeding. In the extreme case of inbreeding, a mother *knows* that all her daughters will be fertilized by her sons. The best sex ratio in this instance is to produce just enough sons to fertilize the daughters, since any other males will be wasted.

The crucial difference between this and the earlier argument for a 1:1 sex ratio is that here the ratio of males to females in the rest of the population does not matter. A female biassed ratio within a brood will not give other parents a chance to benefit by concentrating on sons. An example which supports this prediction is the viviparous mite *Acarophenox* which has a brood of one son and up to 20 daughters. The male mates with his sisters inside the mother and dies before he is born. *Jack Werren* (1980) has tested the prediction that the degree of bias depends on the extent of local mate competition. He studied the parasitoid wasp *Nasonia vitripennis* which lays its eggs inside the pupae of flies such as *Sarcophaga bullata*. If one female parasitizes a pupa, her daughters are all fertilized by her sons and as predicted the sex ratio of her clutch of eggs is biassed towards females.

Only 8.7 per cent of the brood is male. If a second female lays her eggs in the same pupa, what should her sex ratio be? If she lays few eggs she should produce mainly sons, since the first female has laid predominantly female eggs. But as the proportion of the total number of eggs in the pupa that come from the second female will compete for mates also increases. Therefore, her brood should have a female biassed sex ratio. Werren found exactly this pattern: when the second female's clutch was 1/10 the size of the first female's it first female's clutch was 1/10 the size of the first female's it contained only males, but when it was twice as large as he first female's it contained only 10 per cent males, and the quantitative details of the change in sex ratio with relative brood size were much as predicted.

Local Resource Competition or Enhancement

Anne Clark found that the South African prosimian *Galago crassicaudatus* has a male biassed investment ratio among its offspring. She pointed out that this could be explained by the species' life history. As with most mammals, female *Galago* disperse less far than males, and often end up competing both with their mother and with each other for rich sources of food such as gum and fruit trees I the mother's home range. This local resource competition among females reduces their value as offspring: in the extreme case only one daughter might be able to survive on the food available near home, and so investment in other daughters would be wasted. Exactly the opposite effect could arise if the sex that stays at home, rather than hindering one another or their parents, actually helps. In some bird species, males but not females stay at home and help. The consequence of this is to make males slightly more valuable than females as an investment (since they help the parent in its future reproduction) and hence a male-biassed investment ratio might be expected (*Emlen et al.* 1986).

Maternal Condition

It has been stated already that male red deer compete for females by prolonged roaring and antler wrestling contests. In these contests it is an advantage for a male to be big, and size depends among other things on how well fed the male was as a youngster, which in turn depends on his mother's ability to compete for good sources of food and hence produce a plentiful supply of milk. In other words, there is a direct link between a mother's competitive ability while lactating and her son's expected reproductive success. Now if a mother *knew* that her sons would be highly successful harem holders it would pay her to invest heavily in sons rather than daughters: the pay-off in terms of grandchildren would be much greater. Similarly, a mother who *knew* that her sons would not grow up to be big and strong would do better to have daughters, since a daughter's future reproductive success does not depend so much on her mother's milk.

Exactly this pattern has been found in red deer: dominant females who are able to gain access to good feeding sites while lactating and hence produce strapping sons. Tend to have sons, while subordinate females have daughters. It is not known how the sex ratio of adjusted by the mother in the red deer or the *galagos* studied by *Clark,* but the fact that they are adjusted is striking because agricultural geneticists have failed to select for adjustment of the say ratio or to separate male and female sperm of domestic mammals (imagine the value of a female biassed sex ratio to the milk farmer!), and it has often been concluded that mammalian sex ratios are very inflexible. Adjustment of the sex ratio in hymenoptera such as the wasps studies by Warren is not at all a problem because the mother can determine whether an egg becomes male or female by whether or not she fertilizes it.

Population Sex Ratio

When the population ratio of investment deviates from 1:1, compensatory b'as in favour of the rarer sex should occur. In Metcalf's study of *P. metricus* he found that some nests produced only male offspring. These offspring are the product of unfertilized eggs and are produced by workers in nests where the queen has died. In the remainder of the nests in the population *Metcalf* found a female biassed sex ratio, so that the ratio of investment in the population as a whole is 1:1. Finally, it is worth pointing out that the theory of sex ratios discussed here is one example of a more general theory of *sex allocation.*

Other examples of the problem of allocation of resources to male and female reproduction include the division of resources into eggs and sperms by simultaneous hermaphrodites and the timing of sex change in sequential hermaphrodites. The combination of females investing more than males and 1:1 population sex ratio means that males usually compete for females. The potential pay-off for male success is high, so selection for male ability to acquire matings is very strong. Selection for traits which are solely concerned with increasing mating success is usually referred to as *sexual selection.* It can work in two ways: by favouring the ability of one sex (usually males) to compete directly with one another for fertilizations, for example by fighting (*intra-sexual selection*). Or by favouring traits in one sex which attract the other (*Inter-sexual selection).* Often the two kinds of selection act at the same time. The intensity of sexual selection depends on the degree of the difference in parental effort between the sexes and the ratio of males to females available for mating at any one time (referred to as the operational sex ratio). When parental effort is more or less equal, as for example in monogamous birds where both male and female feed

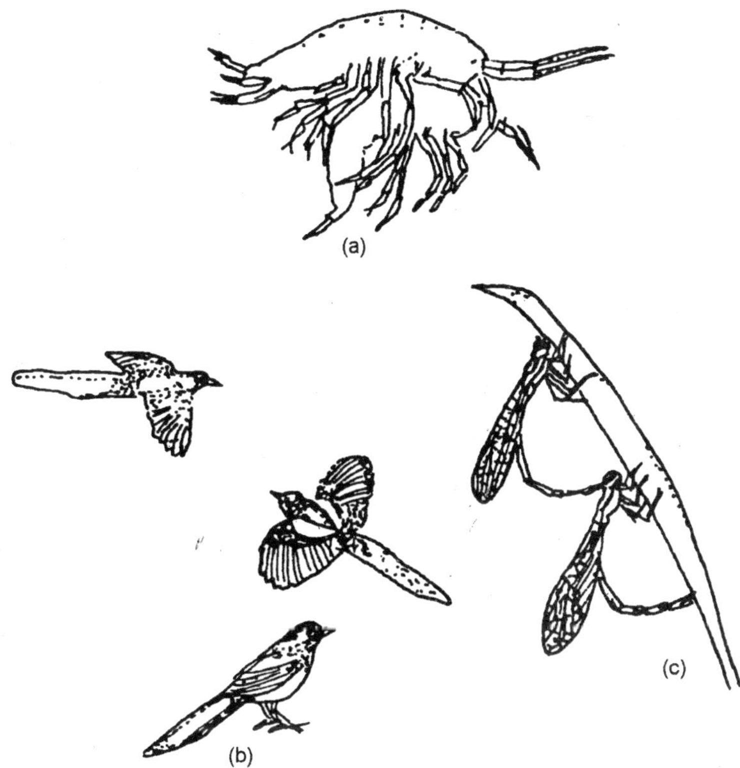

Fig. 10.2. Mate guarding as a form of sexual competition. (a) Precopulatory mate guarding in the freshwater amphipod Gammarus. (b) Male European magpies (Pica pica) assiduously guard their mates against intruding males just before and during the period of egg lying. (c) After copulation the males damsefly guards the female while she lays her eggs, by clasping her thorax with the tip of his abdomen in the 'tandem' position.

the young, sexual selection is less intense than in species will very different levels of parental effort. This follows from the point made earlier that the sex making little investment competes for the sex which makes a large investment.

If equal numbers of the two sexes come into breeding condition at the same time, the degree of sexual selection is reduced because there is less chance for a few males to control access to very large numbers of females. In contrast, when females come into breeding condition asynchronously there is a chance for a small number of males to control many females one after the other. With such high potential pay-offs, sexual competition is very intense.

Ardent Males

The most dramatic and obvious way in which males compete for mates is by fighting and ritualized contests, and often males have evolved weapons for fighting. Males may dispute over direct access to females or over places where females are likely to go, as for example, when male specked wood butterflies defend patches of sunlight. Fighting is often a risky business, as illustrated by the injuries sustained by red deer stages. The most intense fights in many species occur when females are ready to be fertilized and once a male finds a female he often guards her. Males often compete in ways which are less conspicuous than fights, but are no less effective and often more bizarre. The invertebrates are a particularly rich seam of examples.

Female dragonflies, as with many other insects mate with a number of males and store the sperm in a special sac (the spermatheca) in the body for use at a later date. The males compete for fertilizations by trying to ensure that previous sperm is not used by the female. The penis of a male *Ortheturm cancellatum* is equipped with a barbed whip at the end which is used to scrape out of the female any sperm left by previous males before he injects new sperm into the sperm previous males before he injects new sperm into the sperm sac. *Croethemis erythraea*, another dragonfly, uses an inflatable penis with a horn-time appendage to pack the sperm of previous males into corners of the spermetheca. In some invertebrates (especially insects) the male cements up the female's genital opening after copulation to prevent other males from fertilizing her.

The males of *Moniliformes dubius*, parasitic acanthocephalan worm in the intestine of rats, produces a chastity belt of this kind but in addition to sealing up the female after copulation, the male sometimes 'copulates' with rival males and applies cement to their genital region to prevent them from mating again. No less remarkable are the habits of the hemipteran insect *Xylocoris maculipennis*. In normal copulation of the species the male simply pierces the body wall of the female and injects sperm, which then swim around inside the female and injects sperm, which then swim around inside the female until they encounter and fertilize her eggs. As with the acanthocephalan worms, males sometimes engage in homosexual *copulation*. A male *Xylocoris* may inject his sperm into a rival male. The sperms then swim inside the body to the victim's testes, where they wait to be passed on to a female next time the victim mates. Competition between males to prevent each other's sperm from fertilizing eggs is sometimes referred to as 'sperm competition'. The sperm of a second male displaces that of the first male to mate with a female dungfly, Sperm competition also occurs in vertebrates.

For example, during courtship male salamanders and newts deposit little sperm-capped rods of jelly (Spermatophores) on the bottom of the pond and then try to manouever the female onto the spermatophore to achieve fertilization. In the salamander *Ambystoma maculatum* males compete by depositing their spermatophores on top of those of other males. The top spermatophore is the one that fertilizes the female's eggs. A fourth example of the arcane methods of male-male competition found among invertebrates is the use of anti-aphrodisiac smells. *Larry Gilbert* noticed that female *Heliconius erato* butterflies always smell peculiar after they have mated. He was able to show experimentally that the scent does not come from the female herself, but is deposited by the male at the end of mating. Gilbert also found that the scent discourages other males from mating with the female, perhaps because it resembles a scent used by males to repel one another in other contexts.

Reluctant Females

Since females in the great majority of species are the chief provides of resources for the zygote, they might be expected choose their mates carefully in order to get something in return. To put it another way, each egg represents a relatively large proportion of a female's lifetime production of gametes when compared with a sperm, so the female has more to lose if something goes wrong. Mating with the wrong species could cost a female frog her whole year's supply of eggs, but would cost the male very little apart from lost time - he could still go on to mate successfully with a member of the correct species the next day. Not surprisingly therefore, females are on the whole choosier than males during courtship. Choosiness extends not only to discriminating between species, but also to discriminating between males within a species. Females often select males on the basis of material resources they can offer and perhaps sometimes to obtain good genes for their offspring.

Good resources and parental ability

In many animals species males defend breeding territories containing resources which play a crucial role in the survival of a female's eggs or young. For example male North American bullfrogs (*Rana catesbiana*) defend territories in ponds and small lakes where females prefer. One factor which has an important influence on survival of eggs is prediction by leeches (Macrobdella decora). Two environmental features of a territory influence leech predation: if the water is warm the eggs develop faster and are therefore exposed to predation for fewer days, and if the vegetation in the water is not too dense the eggs can form into a ball which the leeches find hard to attack. In territories with a dense mat of vegetation the eggs lie in a thin film on top of the plants and are more easily attacked. The bullfrogs also show that female choice and male-male competition may go hand in hand. The preferred territories are hotly contested by males and the largest, strongest frogs end up in the best sites. Food is a resource which often limits a female's capacity to make eggs and during courtship females may choose whether or not to mate with a male on the basis of his ability to provide food.

In some birds and insects for example, males may provide food for the female during courtship ('courtship feeding') which makes a significant contribution to her eggs. Female hanging files (*Hylobittacus apicalis*) will mate with a male only if he provides a large insect for her to eat during copulation. The larger the insect, the longer the male is allowed to copulate and the more eggs he fertilizes. The female gains from a large insect by having more food to put into her eggs. In birds, the male usually helps to feed the young and courtship feeding may play the additional role of indicating to the female how good the male is at bring food for the young. In the common tern (*Sterna fuscata*) there is a correlation between the ability of the male to bring food during courtship feeding and his ability of the male to bring food during courtship feeding and his ability to feed the chicks later in the season. Pairs often break up during the courtship feeding period and it is possible that females are assessing their mates and rejecting poor quality partners.

Good genes

If some males have better genes than others, could a female improve the success of her progeny by choosing males with good genes? Good genes are ones which increase the ability of her offspring to survive, compete and reproduce. One of the very few studies in which this idea has been tested is an experiment by *Linda Partridge* (1980). She took groups of female fruit flies (*Drosophila*) and either allowed them to mate freely with a population of males or forced each female to mate with a randomly chosen partner. The offspring of the 'choice' and no chose females were then tested for their competitive ability by rearing the larvae in bottles with a fixed number of standard competitors

Fig. 10.3. The male satin bower bird (Ptilonorhynchus violaceus) (left) builds a bower consisting of two parallel rows of vertical fine twigs on a court of cleared ground. The male displays and attracts females to the bower and mating takes place in or by the bower.

(these were distinguishable by a genetic marker). *Partridge* found that the offspring of the *choice* group did slightly but consistently better than those of *no choice* females in the larval competition experiments.

This experiment suggests that females are able to increase the survival of their offspring by choosing good genes in their mates, but it must be born in mind that the results could also be in part explained by intrasexual competition: in the choice experiment, the males that mated may have been superior competitors against other males. The theory of sexual selection is most famous as an attempt to explain the evolution of excessively elaborate adornments and displays of male peacocks, pheasants, birds of paradise and so on. Some elaborate displays may have evolved for use in contests between males, but the most widely accepted (and still controversial) account of how such traits evolve depends on selection by females for good genes.

In this case, however, the genese are not ones which increase the ability of a female's offspring to survive and compete (in fact the bizarre adornments usually seem to be a hindrance to survival) but rather for genes which increase the sexual attractiveness of her sons. *R.A. Fisher* was the first to clearly formulate the idea that elaborate male plumage may be sexually selected simple because it makes males attractive to females. This may sound circular, and indeed it is, but that is the elegance of Fisher's argument. At the beginning, he supposed, females preferred a particular male trait (let us take long tails as an example) because it indicated something about male quality. Perhaps males with longer tails were better at flying and therefore collecting food or avoiding predators. An alternative starting point is to suppose that longer tails were simply easier for females to detect and that they therefore attracted more mates. For example, male natterjack toads *Bufo calamita* with louder croaks attract more females simply because females locate males by moving passively down because females locate males by moving passively down sound gradients. If there is some genetic basis for differences between males in tail length the advantage will be passed on to the female's sons.

At the same time, a gene which causes females to prefer longer than average tails will also be favoured since these females will have sons better able to fly or more readily detected by potential mates. Now once the female preference for loner tails starts to spread, longer tailed males will gain a double advantage: they will be better at flying and be more likely to get a mate. The females similarly gets a double advantage from choosing: she will have sons that are both good fliers and attractive to females. As the positive feedback between female preference and longer tails develops, gradually the benefit of attractive sons will become the more important reason for female choice, and the favoured trait might eventually decrease the survival ability of males. When the decrease in survival counter-balances sexual attractiveness, selection for increasing tail length will grind to a halt. Fisher's verbal argument has been formalized and extended in population genetics models by *Russell Lande.* His analysis underlines the importance of the assumption, implicit in Fisher's argument, that the genes causing variation in preference and in tail length become associated with one another, or in other words show covariance. Although 'tail length' is the classic textbook trait in discussions of sexual selection it is in fact only recently that evidence for sexual selection of tail length in real life has been collected.

Malte Anderson showed that females of the long-tailed widow bird *Euplectes progne* in Kenya prefer males with long tails. This highly polygynous species is an ideal candidate for sexual selection; the male is a sparrow-sized bird with a tail up to 50 cm long. The female's tail is about 7 cm long, presumably close to the optimum for flight purposes. *Anderson* studies 36 males which he divided into four group.

In one group he docked the tails to about 14 cm, while in another group he attached the severed bits of group 1 tails with Super glue. This increased the tail length of group 2 males by an average of 25 cm. The remaining two groups were controls: one lot were left untouched and the others had their tails cut and glued without altering the length. By counting the number of nests in each territory, *Anderson* showed that before his experimental manipulations there was no difference in mating success of the different groups, while afterwards the long tailed males did significantly better than the controls or the shorter tailed birds. Another nice experimental study of a sexually selected elaborate display is that of *Clive Catchpole* on the song of the European sedge warbler.

The song consists of a long stream of almost endlessly varying trills, whistles and buzzes and is sung by the male after arriving back on the breeding territory from the winter quarters: as soon as a male pairs, it stops singing. Catchpole's measurements showed that the males with the most elaborate songs are the first to acquire mates, as expected from Fisher's hypothesis of sexual attractiveness. Further, when female warblers were brought into the laboratory and treated with oestradiol to make them sexually active, they were more responsive to large than to small repertoires. In contrast to the long tail of the window bird it is not obvious what might be the counter-selection limiting the elaboration of songs; one possibility is that elaborate songs are more readily detected by predators, as been found in the leopard frog. Perhaps the most controversial aspect of the idea that females choose males for good genes is the problem of genetic variance. This can be illustrated with an hypothetical example.

Suppose, a farmer wants to select for larger body size in a population of turkeys. He take the heaviest males and females to start the next brood and repeats this procedure for several generations. What will happen? Assuming that there is some genetic variance for body weight, selection will at first be fairly effective, but soon the stock will become less variable with respect to genes for body weight, because only a few genotypes (the heaviest) have been allowed to breed. When the genetic variance is used up selection will cease to be effective in changing body size. In the same way, females cannot improve the genetic quality of their offspring indefinitely by choosing males for good genes. So there is something of a paradox. Fitness-related traits might well be maintained in natural populations either because of new mutations or, more likely, because the optimal genotype is different in different places or at different times so that selection does not favour the same genes for long enough to exhaust all the genetic variation, or migration between populations introduces new variation. *Bill Hamilton* and *Marlene Zuk* have suggested that one kind of selection in particular may have he desired property of favouring different genotypes of different times: selection for resistance to disease. The point can be brought home with a simple hypothetical example. Suppose that a disease-causing organism occurs in two genotypes X and x. Pathogens of type X can infect hosts of genotype F but not f, and similarly pathogens of type x can infect f but not F.

In other words there is a gene for-gene correspondence between pathogenicity and susceptibility to the disease. Now let us start with X as the common form of the disease and F as the common host genotype. Because of its resistance to X, f is at a strong selective advantage and will become commoner, but as f increases in frequency so disease type x gains an advantage relative to X. This in turn swings the pendulum back in favour of host F and thence pathogen X, and so on. It is easy to imagine (and imagination has been made respectable by computer simulations) that the host-pathogen interaction will produce endless cycles of genetic change with constant pressure on both sides to keep ahead or catch up with the other. What does this have to do with sexual selection? The idea is that females might select males on the selection? The idea is that females might select males on the basis of their resistance to disease, resistance which would be passed on to progeny and help in the never-ending race to keep ahead. As Zuk (1984) points out agriculturalists and poultry breeders can readily tell a sic bird by looking at its plumage (for example, a turkey suffering from erysipelas has a swollen purple snood)), so why should females have not evolved a similar ability? Seen in this light, courtship displays and their associated body structures are a sort of medical examination by which one partner assesses the other. As yet there is no evidence that females of any species are skilled diagnostic veterinarians, but Hamilton and Zuk have some evidence that is consistent with their idea.

They argue that sexual selection for external features indicating vigour and disease resistance will have been strongest in those species that are exposed to the most disease (simply because in these species it pays the female to scrutinize her partner closely). Therefore, if traits like he showy plumage and elaborate courtship songs of male passerine birds have evolved as indicators of disease resistance, these traits should be most highly developed in the most diseased species. In a survey of the blood parasites, plumage and song of 109 North American passerine bird species this is generally what Hamilton and Zuk found.

Handicap Principle

Amotz Zahavi has suggested an alternative view of elaborate male displays. He points out that the long tail of the peacock is a handicap in day to day survival, a fact which few would dispute. He then goes on to suggest that longer tails are preferred by females precisely because they are a handicap and therefore act as an advertisement of male quality. The tail demonstrates a male's ability to survive in spite of the handicap, which means that he must be extra good in other respects and is able to pass on any genetic component of this quality to his sons and daughters. The difficulty with this argument is that the offspring inherit not only the good genes but also the handicap, and under most conditions they will be better off without either than with both.

For this reasons, Zahavi's idea is unlikely to work in its original form, but if we modify it to say that the handicap is not a fixed trait but is simply conditional upon male vigour its major problem disappears. In this modified version, the size of a male's handicap is a direct index of his current vigour, as in Hamilton and Zuk's parasite hypothesis where a male's plumage or other displays indicate his degree of disease resistance. In fact, the idea that the intensity of a male's sexual displays are not genetically fixed (as in Fisher's hypothesis) but are simply a phenotypic expression of his current condition, can be extended beyond the handicap principle. Any display or trait (antler length, croaking rate, and so on) whether or not it is a handicap, could be a reliable index of a male's well-being and be used in assessment by females or rival males.

Male Investment

We have so far assumed that females are investors and males are competitors. While this picture describes most animal species, there are exceptions. In many birds, some amphibians, and arthropods, both male and female invest about equally in the eggs or young by feeding, guarding or brooding. Sometimes the usual sex roles are completely reversed so that males do the investing and females the competing. The ideas about sexual conflict and sexual selection can still be applied in modified form to species with equal or primarily male investment. When both sexes care equally for the offspring, for example, courtship may involve assessment and choice by males as well as by females.

Males of species with internal fertilization can never be absolutely sure that they have fathered the children of their male, and one role of courtship may be as an insurance against cuckoldry.

A prediction of this idea is that courtship allows males to assess whether or not females have previously mated with others. This was tested by Erickson and Zenone. They found that male barbary doves (*Streptopelia risoria*) attack a female instead of courting her if she performs the bow posture (an advanced stage of courtship) too quickly. Since the females which responded in this way had been pretreated by allowing them to court with another male, the reaction of the test males in rejecting eager females is adaptive if courtship plays a role in assessing certainty of paternity, before investing in offspring. It would not have been predicted by the older view that male courtship serves to sexually arouse the female! In species with high male investment, females tend to be the competitive sex and males may be choosy.

In moorhens (*Gallinula chloropus*) males do almost three-quarters of the incubation and females play an active role in competing for the chance to mate with good incubators. These ideal husbands are small and fat: well equipped to survive on their reserves during long incubation stints. In other species, investing males may actually reject low quality females.

Sexual Conflict

Recall the view of the origin of anisogamy as the primaeval example of sexual conflict. The conflict was one about mating decisions. Macrogametes might have done better had they been able to discriminate against microgametes, but in the evolutionary race microgametes won. Similar, but more directly observable, conflicts by interest between the sexes are still apparent today, not only with respect to mating decisions but also in the contexts of parental investment, multiple matings and infanticide.

Mating decisions

As we have emphasized, females have more to lose and therefore tend to be choosier than males. Thus for a given encounter it will often be the case that males are favoured if they do mate and females if they do not. An extreme manifestation of this conflict is enforced copulation as exemplified by scorpionflies (*Panorpa* spp.). Male scorpionflies usually acquire a mate by presenting her with a nuptial gift in the form of a special salivary secretion or a dead insect (this is very similar to the Hylobittacus described earlier). The female feeds on the gift during copulation and turns the food into eggs. However, sometimes a male enforces copulation: he grasps her with a special abdominal organ (the notal organ) without offering a gift. Enforced copulation appears to be a case of sexual conflict.

The female loses because she obtains no food for her eggs and has to search for food herself, while the male benefits because he avoids the risky business of finding a nuptial gift. Scorpionflies feed on insects in spider's webs and quite often get caught in the web themselves, so foraging is certainly risky (65 per cent of adults die this way). Why do not all males enforce copulations? The exact balance of costs and benefits is not known, but it appears that it results in a very low success rate in fertilizing females, so perhaps males adopt this strategy only when they cannot find prey or make enough saliva to attract a female.

Parental investment

It is sufficient to note that in species with investment beyond the gamete stage, each sex might be expected exploit the other by reducing its own share in the investment. The outcome of this sexual conflict may depend on practical considerations such as which sex is the first to be in a position to desert the other. When fertilization is internal, for example, a male has the possibility of deserting the female immediately after fertilization and leaving her with the eggs or young to care for.

Infanticide

Male lions may slaughter the cubs in a pride shortly after they take over as group leaders. This behaviour (Which is also seen in some primates) probably increases male reproductive success, and clearly decreases female success. This seems to be a case of sexual conflict in which the males have won, but it is perhaps surprising that females have not evolved counter-adaptations. They could, for example eat their own young once they have been killed in order to recoup as much as possible of their losses.

Multiple matings

As Bateman's experiments with *Drosophila* showed females often gain little by mating with more than one male. However, because of sperm competition males may gain by mating with already fertilized females. Multiple matings are likely to be costly to the female at the same time as being advantageous to the male. This dramatically illustrated by he dungflies. When two males struggle for possession of a female, the female is sometimes drowned in cowdung by the fighting males on top of her! Conflicts of interest between the sexes will lead to an evolutionary rate of the sort envisaged by Parker et al. for sperms and eggs. There is no simple answer to the questions 'Which sex is more likely to win the chase?" As we discussed earlier, factors such as the strength of selection and the amount of genetic variation will determine how fast the two sexes can evolve adaptation and counter-adaptation, but it is not possible to make any more specific statements about the outcome of sexual conflict races.

Significance of Courtship

Some aspects of courtship behaviour can be interpreted in terms of sexual conflict and sexual selection. However, this is not true of all courtship signals; many are designed for species identification, and here the interests of the two sexes are similar because both benefit by mating with a member of the same species. Some of the clearest examples of this role of courtship come from studies of frog calls. When several species of frogs live in the same pond, each has a characteristic and distinct mating call given by the male, and females are attracted only to calls of their own species. In some frogs (e.g., the cricket frog *Acris crepitans*) it has been shown that the female's selectivity of response results from the fact that the auditory system is turned to the particular frequencies in the male call. Courtship displays may also pay a role in competition between males within a species for mating opportunities.

Often the same displays simultaneously serve to repel other males and attract females. An example for which this has been demonstrated experimentally is the mating call of the pacific tree frog (*Hyla regilla*). Males are repelled and females attracted by loudspeakers broadcasting the mating call and females select out of a group of loudspeakers the one which calls for the longest bouts. Females may choose between displays purely on the basis of sexual attractiveness, as explained by Fisher's theory of sexual selection, but there is also the possibility that differences in courtship between males may indicate habitat quality, for example males with territories containing a lot of food might be able to afford to spend more time displaying. A third role of courtship to which we have referred is assessment.

In a species with male parental care females may assess the ability of the male to look after young and males may assess whether a female has previously been fertilized. Early work by ethologists on birds and fish showed that at the beginning of courtship males are often aggressive and females are coy or reluctant. Courtship was seen, therefore, as serving to synchronize sexual arousal of the partners. A possible explanation of why it should be necessary to overcome aggression and reluctance is that the early phases involve assessment by both partners before investing in offspring.

Mating Systems

Despite the argument, bases upon relative parental investment by the sexes, that polygyny should be the normal mating system in animals, monogamy is common in some groups, for example, in 90% of birds. Parental investment in many birds, however, is almost as great in the males as it is in the female. Young altricial birds in nests are very vulnerable to predation and survive best if they can develop rapidly. If the female alone can seldom rear two young, it is advantageous for males to help with parental care. *Armstrong* (1955) suggested that wrens (*Troglodytes troglodytes*) are monogamous, when both parents are needed to collect enough food for the young to develop. In long-lived birds, long-term mate fidelity seems to result in great efficiency of reproduction due to rapid mating and effective co-operation between mates (*Coulson* 1966, *Davis* 1976). *Crook* (1964), who compared 90 species of weaver bird, *Ploccinae*, found that the primarily insectivorous species which inhabit forests are monogamous whereas savannah–dwelling seed-eaters are colonial and polygynous.

The monogamous species defend and forage in territories while the seed-eater living in flocks and probably obtain food more readily hence the males do not have to feed the young. Habitat quality may affect mating systems. Verner (1964) found that female long-billed marsh wrens (*Telmatodytes palustris*) would mate with a male which was already mated, rather than with an unmated male, if the quality of the territory of the first male was much better than that of the second. *Verner* and *Willson* (1966) therefore proposed that there is a polygyny threshold in habitat quality. This idea was developed by *Orians* (1969) who quotes supporting evidence from his own studies of red-winged *Agelaius phoeniceus* and yellow headed *Xanthocephalus xanthocephalus* black birds. Orians proposes that polygyny should be commoner in habitats where food is locally abundant, especially if the number of nest-sites is restricted and that it should be prevented in precocial birds.

In such situations, genes which have the effect of favouring polygyny if the male himself, or the territory which he controls, is of high enough quality, should spread in the population. It seems that this has happened in habitats where there is a period of rapid resource increase. In addition to marsh and savannah dwellers, such as those mentioned, polygyny occurs in many upland game birds, in the ruff *Philomachus pugnax* and in a few sandpipers. *Emlen* and *Oring* (1977) summarise the conditions necessary for the occurrence of polygamy in terms of the relative economics of defending one or several mates. This depends upon the spatial distribution of resources, the sex ratio and the time at which mates become available. If females show synchrony in their sexual receptivity, polygyny is more difficult.

Emlen and *Oring* describe three major types of polygyny, involving resources defence by males, harem defence, or selection by females from male assemblies such as leks. It is nor surprising that polygyny is the normal mating system in mammals for the female's parental investment is always large. *Eisenberg* (1966) lists a few cases of monogamy, for example, marmosets (Callitrichidae), gibbons (Hylobatinae), beavers (Castoridae), one seal and a few terrestrial carnivores. An intermediate between polygyny and polyandry, in which females lay a second batch of eggs which the male incubates or guards but where the male may also mate with other females was called *rapid multiple clutch polygamy* and *Emlen* and *Oring*. This is common in fish and occurs occasionally in birds.

Polyandry has not been reported for mammals (*Eisenberg* 1966) but *Jenni* (1974) describes it in several species of birds. Most are in the order Chondriiformes and all have precocial young. The females of the American jacana (*Jacana spinosa*) defend large territories in march areas which include the smaller territories of several males. The males perform all incubation and parental care but they defend their territories against other males. In addition to such resource defend their territories against other males. In addition to such resource defend polyandry, *Emlen* and *Oring* describe harem polyandry and female access polyandry. One consequence of recent work on mating systems is that is has become apparent that they can vary within species and within individuals according to the conditions which obtain at any time. Emlen and Oring list examples of variation between polygyny and monogamy; from monogamy to rapid multiple clutch polygamy to polyandry, and from lekking to mate or resource defend polygyny. Detailed study of individuals emphasizes the plasticity of mating systems and raises question about the factors which affect reproductive behaviour, both during development and at the time that decision about the behaviour are taken.

Struggle of Males and Females

Because the males reproductive success is probably limited by the number of females they can attract and defend from other males rather than their sperm supply natural selection will favour any property in a male that enables him to mate with more females. This general principle finds its natural realization in the fascinating diversity of adaptations by which the males of different species seek to increase their share of the mating, Let us examine some examples. The most obvious form of competition for females is straightforward fighting. We have considered the subject before, in the previous chapter, and we only need notice here the kinds of characteristics it leads to in males. It will favour strength, and fighting in its simplest form will favour increased size. In toads, for instance, the males sit on the backs of females for a few days before the female lays her eggs. Other males try to dislodge the sitting males from the females, by pulling them off. In experiments larger males are much better at dislodging smaller males from females than vice versa. There is an advantage in being large in males, and large males are most likely to mate with females than are small males. In other species males have evolved special weapons with which they fight over females.

It is found only in males, who use their tusks to fight each other. Most males suffer severe wounds from these fights and in one sample from a narwhal population over 60% of the males had broken tusks. Other strange weaponry can be seen in the males of some kinds of beetles, and the antlers of deer have evolved for the same reason. In species that fight, a male can increase his effective strength by forming a coalition with another male, and ranging up on competitors. Males lions do just that. Groups of two or three male lions try to take over harems of females by forcibly evicting the existing male owners. Bigger coalitions of males are more successful in taking over harems. However, physical fighting is not the only means by which males compete with each other. When a coalition of male lions successfully takes over a harem of females, the first thing the males do is to kill all the young lion cubs in the pride. The cubs, of rouse, were fathered by the previous males, and are no loss to the new owners. There is even a gain to them. While a lioness is lactating for her cub she will not produce another cub; but when her cub is killed (or is weaned) she soon becomes ready to reproduce again. By killing the cubs, the males bring forward the time when they can start reproducing. Infanticide by males which have just taken over a harem is probably common in nature in many

Fig. 10.4. Mating behaviour in Stickle back.

species: it has been observed several times in the Hanuman langur (Presbytis entellus, a species of primate) and has been anecdotally recorded in many other mammals. A more subtle, less gory, form of male competition is the microscopic battle fought among the sperms of different males.

In many species, if a female mates with two males, the second male by some means or other manages to fertilize many more than half the eggs. In the fruitfly *Drosophila*, for example, the second male fertilizes from 83% to 99% of the eggs. Males have evolved other kinds of counter-adaptations to prevent sperm competition. Some male damselflies stay with their mates after copulation and fight off any other males who come near; only after the female has laid her eggs does the male leave her.

Likewise, the acanthocephalan worm *Moniliformis moniliformis* male sticks a 'chastity belt' on a female after mating, probably to prevent other males from copulating with her. The males are also know to 'rape' other males, cementing up the victims' genital openings to render them incapable of copulation. An extraordinary kind of sperm competition has been found in the hemipteran insect *Xylocoris maculipennsi* by J. Carayon. Copulation in this species is achieved by injection, the male simply punctures the side of the female with his genitalia and squirts his sperm in. However, a male sometimes compulates with another males. His sperms then migrate to the victims tests, and when the second male comes to compulate with a female, the first male's sperm will be injected in her.

Courtship

Courtship behaviour functions to bring together two animals of different sexes of the same species under conditions where mating is likely to occur and to be successful. The first problem is that of simply locating a potential mate. Obviously, the conspecific must be of the opposite sex if reproduction is to be successful. Courtship often entails a complex sequence of interacting signals, that function to ensure that an animal mates with an appropriate partner. Timing is an important part of successful reproduction; both male and female must be in appropriate physiological condition for mating. This is ensured by the synchronization of cycles that results from the interactions of environmental stimuli and of the two animals themselves. The classical ethologists pointed out that many courtship patterns contain elements of conflict, often because the initial response of an individual to a stranger in its vicinity may be aggressive. Mating systems vary greatly from species to species. Some species, such as some swans and geese, are truly monogamous and pair for life.

Many migrant birds from pair bonds that last for a single season. Some primates are serially polygamous, in that they pair with several individual partners, each for a deniable period of time. In simultaneous polygamy, an individual maintain simultaneous pair bonds with several individuals of the opposite sex. Many mammalian species appear to be completely promiscuous, with many copulations and no pair bonds (Brown, 1975).

Arthropods

Initial detection often occurs via of action. For example, silk moths, *Bombyx mori*, are noted for the extreme sensitivity of the male to a sex attractant, bombykol, produced by the female. Males are attracted from great *Drosophila* (*Spieth*, 1974). Various combinations of behavioural patterns-such as wing vibration, leg vibration, wing semaphoring, circling, and licking-make up the courtship patterns of different species.

Fishes

Different specie of fishes display diverse courtship patterns. The behaviour of tropical aquarium fishes has been most carefully described. Blackchin mouthbreeders, *Tilapia melanotheron*, display four "pure" courtship patterns-patterns that occurs in virtually no other contexts. These are the nod (rapid titling down and forward), quiver (a pattern of head shaking), nib (biting the substrate), and skim (remaining motionless near the nest) (Barlow & Green, 1970). The most prominent display in the courtship of guppies is the adoption of a "sigmoid" or S-shaped posture by the male. Male platys approach females with a sidling movement and also display patterns of backing toward the female, quivering of the body and swinging of the gonopodium (Clark, Aronson, and Gordon, 1954).

Amphibians and Reptiles

Male bullfrogs adopt territories, from which they emit their familiar loud choruses. Females appear attracted by such choruses. Many crocodilian species emit loud roads. The courtship behaviour of anoles (*American chameleons*), *Anolis carolinensis*, is familiar to many Americans with backyards. The male displays a rhythmical up-and-down bobbing of his body coordinated with the exposure of his dewlap, a bright red flap of skin beneath the chin. Courtship in mot species of snakes is based on tacttie stimulation of the female and olfactory stimulation of the male. The dramatic so called "courtship dances" in which two snakes become closely interwined have generally proved to be male-male interactions, presumably aggressive in nature.

Birds

Birds have provided some of the most dramatic examples of courtship behaviour. The complex displays of grebes, gulls, ducks, herons, and other birds have been favoured subjects of study by ethologists. The attention paid to bird songs and the legends surrounding them is indeed substantial. Lorenz studies various courtship patterns of male ducks. These can be seen in many duck pons. In the grunt-whistle, the male lowers its bill to the water, arching the body upward, flicking the bill, and emitting a loud whistle followed by a grunting sound.

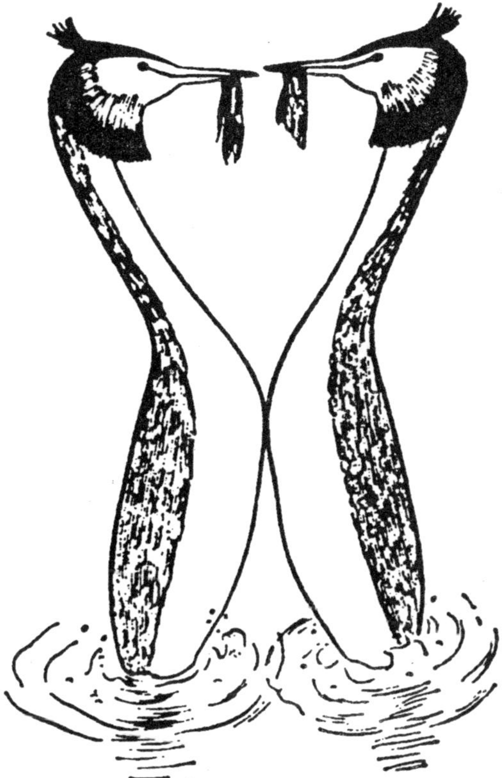

Fig. 10.5. Male and female great crested grebes, Podiceps cristatus, performing one of their elaborate mutual displays.

The head-up-tail-down display in accompanied by a bound whistle. In the down-up display, the breast is dipped into the water and the bill is jerked upward and outward, flipping a column of water as it goes. In green herons males set up territories. The full-forward display functions to drive away other males from a resident male's territory. The calls of the males attract females. Females are initially repelled with full-forward displays, but females are initially repelled with full-forward displays, but females persist and courtship is eventually initiated with the appearance of the snap display and the stretch display. After pairing, the two birds fly about the territory, occasionally showing the more intense flap-flight display. Display patterns then become more contact-oriented until copulation occurs. Many birdsongs function to repel males from territories and to attract females. Song patterns are as characteristic of some species as are their appearance.

Fig. 10.6. Courtship preening in four species of duck; in every case the movement serves to emphasize bright marking on the wings.

Mammals

Ewer (1968) surveyed some courtship patterns of mammals. Olfaction plays a large role in the regulation of courtship in many species. Patterns of anogenital investigation and urine testing are common. After smelling a female, the males of many species display a *Flehmen* response, in which the neck is extended and the upper lip is curled. This pattern appears to function more in facilitating perception of the odor than as a display. Female mammals often solicit mountings, sometimes by approaching a male, nuzzling or licking him, and often by darting from him. Much of the running from the male done by females appears more as solicitation behaviour than escape. Courtship in bottle-nosed dolphins entails vocalization, mouthing of the partner, nuzzling of the partner's genitalia, rubbing of bodies, stroking with flukes or flippers, displaying of the white underside, leaping chasing and head butting (*Puente* and *Dewsbury*, 1976).

Mating

The consummation of courtship activity is the occurrence of actual mating, wherein fertilization of ova is accomplished. Patterns of mating are almost as diverse as are courtship patterns, including both internal and external fertilization via a great variety of process. A select sample of mating patterns will be described.

Arthropods

In many species, sperm are "packaged" in a spermatophore, a kind of bag or sack containing the sperm. Perhaps the heights of impersonality are reached in some species of mites, pseudoscorpions, millipedes, and springtails, in which the male deposits a spermatophore on the substrate and the female comes along later to pick up the package. The pair may never meet. Many aquatic species have external fertilization (e.g., the male horseshoe crab, Limulus, hooks himself on behind the female and remains there until eggs are released). Most terrestrial forms have some kind of internal fertilization, wherein the spermatophore is placed directly into the female using some kind of appendage.

Fishes

Fertilization in fishes can be either internal or external. In anabantid fish, such a gouramis and betas, a bubble nest is built and spawning generally takes place beneath it. Eggs and sperm are released simultaneously and eggs are fertilized as they float up to the nest. In other species – such as guppies, platys, and *Gambusia* – a gonopodium (modified anal fin) is used to transfer sperm to the female. Male nurse sharks grasp the hind edge of one of the female's pectoral fins, turning her on her back, and inset their specialized claspers.

Amphibians and Reptiles

Most species of salamanders accomplish internal fertilization through use of spermatophore, although external fertilization and parthenogenesis occur in some species. Most frogs and toads have external fertilization, with eggs fertilized as they re laid. Exceptions include tailed frogs, which have an intermittent, organ, and African toads, in which cloacal glands of the male and female are brought together. Typically, a ripe female frog orients toward a calling male and is clasped by him in a state termed "amplexus." Eggs are released and fertilized in several bouts. If a male clasps another male or is clasping a female that has finished spawning, a release call by the recipient results in release by the clasping male (*Rabb,* 1973). Reptiles display internal fertilization through use of intromittent organs. Among anoles, the male delivers a neck grip, twists his tail around that of the female, and inserts his hemipenis.

Birds

In contrast to their courtship activities, the copulations birds generally show little variability. In most species there is no penis, and thus sperm are transferred from the cloacal gland of the male to that of the female. In copulation, the two cloacal glands are brought together in the "cloacal kiss." In Japanese quail, for example, the male grasps the feathers of the female's neck in his break, mounts her by standing on her back, and orients until the cloacal glands can be brought together.

Mammals

Much of the research on mammalian copulatory behaviour has been done using laboratory rats, *Rattus norvegicus*, as subjects. Their copulatory behaviour will be described in some details. If a receptive female is introduced into a cage containing a vigorous male, a predictable sequence of events ensues. After initial courtship behaviour (sniffing, chasing, soliciting by the female), the male pursues the running female. As the male mounts the female from behind, the female adopt a stereotyped posture, termed "lordosis." The head and hindquarters are elevated to produce a concave arching of the back, while the tail is deflected to the side. Two kinds of copulations typically occur. In some, called "intromissions," the male's penis is inserted into the female's vagina for a period of about ¼ second before the male displays a rapid dismount. Although there may be extravaginal pelvic thrusting by the male prior to insertion, there is no intravaginal thrusting.

After about ten such brief intromissions spaced about a minute apart from each other, the male displays a different pattern, called "ejaculation." The male shows a kind of convulsive thrusting during ejaculation and clasps the female for a few seconds before the dismounts (in a stereotyped vertical posture). A complete group of intromissions ending with an ejaculation is termed a "series." Ejaculation and the associated sperm transfer never occurs without prior intromissions. Ejaculation is followed by a temporary cessation of sexual activity. After about 5 minutes, activity resumes and a second series occurs. Normally, rats will complete about seven ejaculatory series before they may be considered satiated, according to a criterion of ½ hour with no intromissions.

INDEX

Q

R

S